Societal Dimensions of Environmental Science

Societal Dimensions of Environmental Science
Global Case Studies of Collaboration and Transformation

Edited by
Ricardo D. Lopez

CRC Press
Taylor & Francis Group
Boca Raton London New York

CRC Press is an imprint of the
Taylor & Francis Group, an **informa** business

CRC Press
Taylor & Francis Group
6000 Broken Sound Parkway NW, Suite 300
Boca Raton, FL 33487-2742

First issued in paperback 2020

© 2019 by Taylor & Francis Group, LLC
CRC Press is an imprint of Taylor & Francis Group, an Informa business

No claim to original U.S. Government works

ISBN 13: 978-0-367-67055-9 (pbk)
ISBN 13: 978-1-138-05445-5 (hbk)

Library of Congress Cataloging-in-Publication Data

Names: Lopez, Ricardo D., editor.
Title: Societal dimensions of environmental science : global case studies of collaboration and transformation / edited by Ricardo D. Lopez.
Description: First edition. | Boca Raton, FL : CRC Press/Taylor & Francis Group, 2019.
Identifiers: LCCN 2018049602 | ISBN 9781138054455 (hardback : acid-free paper)
Subjects: LCSH: Environmental sciences—Social aspects. | Environmental sciences—Public opinion. | Environmental policy—International cooperation. | Environmental protection—International cooperation. | Globalization—Environmental aspects.
Classification: LCC GE40 .S63 2018 | DDC 304.2—dc23
LC record available at https://lccn.loc.gov/2018049602

Visit the Taylor & Francis Web site at
http://www.taylorandfrancis.com

and the CRC Press Web site at
http://www.crcpress.com

To listen well is as powerful a means of influence as to talk well

and is as essential to all true conversation.

Chinese proverb of unknown attribution

Contents

Preface

A friend of mine is the general manager of an oceanfront hotel in one of the most picturesque coastal areas of the western Pacific—a lovely property with large floor-to-ceiling windows that cover the entire coastal-facing front lobby area and restaurants of the glistening beachfront hotel. As with most hotels, even in paradise, comes the perennial and constant rush of special needs, emergency situations, and pressures of a demanding clientele on holiday. Included in the daily activities of a general manager are mediating late-night arguments at the front desk, domestic disturbances in rooms, and loud/rude patrons, all intermixed with a fair share of complaints in multiple languages, by visitors from across the world. These situations also include juggling the many concerns of the local residents, the police, and other establishment owners in this complex and bustling area of a competitive and commercially driven part of town. My friend, the general manager, often describes to me the scene, which is usually around 2 am, of a lobby of shouting people, all solely intent on their personal needs and complaints.

In such circumstances, one would never expect cooperation to emerge, much less complete coordination and collaborative solutions, and yet, there was a miraculous change in this attitude on a fateful day in early monsoon season, which led to the "community" saving the hotel, quite literally. The stimulus that brought about this miraculous cooperative event was a typhoon, one that eventually took the lives of over 100 people and injured countless more in the seaside region in which the hotel resides. The short version of the story is that the typhoon, with sustained winds in excess of 100 miles/h (160 km/h), brought about tremendous solidarity with the remaining guests, the local residents, and neighboring proprietors, many of whom exhibited such altruism that one would never have guessed they were the same (typically) squabbling group described earlier in this story. The many people involved acted valiantly and, for many hours while the winds were at maximum velocities, prevented the giant glass windows of the gleaming hotel from being blown in. In the end of the story, the lobby and restaurants were saved, there was minimal damage to the large windows, and the hotel reopened after relatively few repairs. True heroism on the part of the many who came to the rescue, locally, compelled to assist by nothing other than an instinct to band together to confront the situation.

Why did this happen—what was this instinct that had arisen? It is this instinct, one that is uncovered at certain pivotal moments in the lives of humans, that can simply be called an imperative, and what I specifically refer to in the coming chapters as a "societal imperative," which drives cooperation among diverse groups, often with divergent interests, in the face of a commonly recognized crisis. In this instance, those in the hotel and those in the neighborhood at the critical time of an approaching storm realized what was at stake, and with little time to debate, they acted with a common goal, which resoundingly had to be achieved. In such circumstances, we all tend to come together, not as solitary or autonomous people but rather as interdependent individuals, much more reliant on each other than we might normally be in other circumstances. This situation is analogous to a number of other societal imperatives, which all humans around the world tend to share and value, such as the following:

- Breathable and pleasant air
- Water that is clean enough to not harm our health and that tastes good
- Secure, safe, or otherwise consumable food supplies
- A safe place for ourselves and family to reside
- A clean and healthful environment to enjoy

This book is about the restoration and the maintenance of these fundamental societal imperatives—some would call them "ecological goods and services" or "ecosystem services." To explore these societal imperatives and how humans value them, this book was developed to focus on "how we do the work" rather than on "what the work is." This book focuses on people's needs and the relationships among people, societal norms, and their relationships with their environment. Specifically, this book describes how people come together to address the societal imperatives in their lives, for it is these "societal dimensions" that are served by the professionals that work in the realm of the environmental sciences. Without the dedication of those who care about addressing the societal imperatives listed earlier, we risk the degradation of basic ecological goods and services that civil societies around the world depend upon, and indeed expect more and more. Let us now celebrate the commitment of some of these very dedicated people, as described in this book, who have devoted themselves to utilizing the methods of collaboration to transform the lives of their fellow community members and the lives of future generations to come.

Ricardo D. Lopez
September 2018

Editor

Dr. Ricardo D. Lopez, PhD, has been a leader in the field of environmental sciences over the past three decades. During his tenure in academia and public service with the Ohio Environmental Protection Agency, the U.S. Environmental Protection Agency, and the US Forest Service, he has led science teams in a variety of geographically diverse areas, from the Asia-Pacific Region to the Americas, focusing on the application of science to meet the diverse needs of communities, with a strong emphasis on public service. He has always had a passion for understanding the needs of everyday people, and bringing knowledge and solutions to bear on those topics, wherever it may be possible.

With his background and expertise in remote sensing and field-based approaches for environmental analyses, Ric has conducted and published on a number of, both, theoretical and applied environmental topics. This body of work includes monitoring and assessing of terrestrial, aquatic, and transitional ecosystems; invasive plant species; multi-scale indicators of sustainability; and solutions to risk-based landscape ecological issues.

A native of California, Ric spent much of his life exploring, appreciating, and writing about the diverse aspects of complex landscapes, from the tropics to temperate regions, bringing his expertise as a landscape ecologist to bear on specific local, regional, and global environmental issues. Ric earned his BS in ecology, behavior, and evolution at the University of California, San Diego, and his master's and doctoral degrees in environmental science at The Ohio State University, with an emphasis on landscape ecology and wetland ecology. Ric is currently the regional director of Partnerships & Collaboration for the US Forest Service's Pacific Southwest Research Station.

Contributors

Bubaker Hamad Almansori
International Institute for Halal
    Research and Training (INHART)
International Islamic University
    Malaysia (IIUM)
Selangor, Malaysia

Marjorie V. Cushing Falanruw
USDA Forest Service, Pacific
    Southwest Research Station
Institute of Pacific Islands Forestry
Yap, Federated States of
    Micronesia

Shawn Grindstaff
Facilitator and Process Consultant
Valparaiso, Indiana
United States of America

Brenda Groskinsky
Environmental Scientist
Lawrence, Kansas
United States of America

R. Flint Hughes
USDA Forest Service, Pacific
    Southwest Research Station
Institute of Pacific Islands Forestry
Hilo, Hawai'i
United States of America

Irwandi Jaswir
International Institute for
    Halal Research and Training
    (INHART)
International Islamic University
    Malaysia (IIUM)
Selangor, Malaysia

Kristin Jayd
Department of Entomology
University of Maryland
College Park, Maryland
United States of America

Mohammed A. Kalkhan
Natural Resources Environmental
    Laboratory
Colorado State University
Fort Collins, Colorado
United States of America

Lloyd L. Loope
USGS Pacific Island Ecosystems
    Research Center
O'ahu, Hawa'i
United States of America

Ricardo D. Lopez
USDA Forest Service
Pacific Southwest Research Station
Albany, California
United States of America

Richard A. MacKenzie
USDA Forest Service, Pacific
    Southwest Research Station
Institute of Pacific Islands Forestry
Hilo, Hawai'i
United States of America

Muhammad Elwathig S. Mirghani
International Institute for Halal
    Research and Training (INHART)
International Islamic University
    Malaysia (IIUM)
Selangor, Malaysia

Maliha S. Nash
System Exposure Division,
    Ecological & Human Community
    Analysis Branch
U.S. Environmental Protection
    Agency
Las Vegas, Nevada

Fitri Octavianti
Faculty of Dentistry
Universiti Sains Islam Malaysia
    (USIM)
Kuala Lumpur
Malaysia

Reed M. Perkins
Department of Environmental
    Science and Chemistry
Queens University of Charlotte
Charlotte, North Carolina
United States of America

Hong Tinh Pham
Faculty of Environment
Hanoi University of Natural
    Resources and Environment
Hanoi, Việt Nam

Francis Ruegorong
Yap State Division of Agriculture
    and Forestry
Yap, Federated States of Micronesia

Ardena Saarinen
Natural Resources & Environmental
    Management
University of Hawai'i Manoa
O'ahu, Hawai'i
United States of America

Sahadev Sharma
Institute of Ocean and Earth Sciences
University of Malaya
Kuala Lumpur, Malaysia

1

The Journey to Integrating Societal Needs and the Environmental Sciences

Ricardo D. Lopez
USDA - Forest Service, Pacific Southwest Research Station

Lloyd L. Loope
USGS Pacific Island Ecosystems Research Center

Ardena Saarinen
University of Hawai'i Manoa

R. Flint Hughes
*USDA - Forest Service, Pacific Southwest Research Station,
Institute of Pacific Islands Forestry*

CONTENTS

1.1 Who Actually Cares about the Societal Dimensions of the Environmental Sciences?

Any environmental scientist, or scientist of any type for that matter, knows that the effectiveness of communication directly corresponds with the success of the real-world application of their research findings. Among the most important indicators of how society will utilize the work that has been conducted by any environmental scientist is the extent to which scientific outcomes are integral to the practical needs of the communities that are to be served by the scientific outputs. Although most scientific research endeavors typically emanate from the theoretical bases of the sciences, the eventual linkages made between the theory and how that information affects people, as well as their perception and understanding of the science and the scientists themselves, are very important to recognize. As one ponders how to measure or gauge the linkages between science and the people affected by the science, some scientists have come to realize how little formal training exists that delves into these "societal dimensions" within the individual subdisciplines of the environmental sciences, and how far they often need to go to fully connect their work with the communities that they are genuinely endeavoring to serve. However, a growing proportion of environmental scientists have been developing methods for specific inquiry and collaboration, which harness robust approaches to assess what has been traditionally referred to as "stakeholder needs," and formalizing methods and tools to quantify what has been traditionally referred to as "science delivery" or "outreach and education" efforts, to people and communities around the world. Contemporary approaches are going well beyond the concepts of "stakeholders" and "delivery" of knowledge, to a less biased and more integrative and collaborative approach to understanding and solving problems in a way that emanates from the needs of communities and people. Such approaches can be controversial, and by no means is there unanimity in these integrative community approaches, and chief among the challenges for utilizing more integrative approaches to collaboration is that these approaches are not always easily quantified or characterized.

To assist you in developing your own more integrative approaches to address the societal dimensions of the environmental sciences, a number of distinguished and experienced practitioners of a diverse set of skills and disciplines have come together in this book to share their experiences, outlined in a series of detailed examples in the coming chapters, all derived from successful applied work in the environmental sciences. This book is an intentionally selected compilation of experiences, which shares a multitude of rich stories from around the globe, each of them with their fair share of successes, as well as challenges along the way toward their societally driven goals. Included in each of these very authentic stories are genuine moments of collaboration and transformation, and primary knowledge, and indeed involvement, of the authors in the circumstances. Be they leaders in the community, key decision makers, environmental scientists, or other important partners, each of the individuals described in this book was driven by a desire to achieve some type of balance between the scientific and other societal imperatives. All of humanity continues to struggle with the pressing environmental issues of our time, as evidenced by the news stories of the day, such as the major topics covered in this book: adequate and safe water, sufficient and affordable food supplies, and the sustainable utilization of natural resources, while respecting and perpetuating cultural norms and maintaining traditions.

The particular global stories in this book were specifically selected to provide important new methods and approaches necessary for informed decision-making and to provide a better understanding of the common threads of learning, collaboration, negotiation, and compromise in the application of the environmental sciences, within the larger context of societal values and needs. The stories in this book go well beyond typical case studies by sharing the key societal process dynamics involved, including the history and traditions of those involved, as well as the relevant science—each story provides you, the reader, with deep lessons for use in a diversity of science applications, geographies, and societal norms and circumstances.

This book will increase the capability of both seasoned professionals and apprentices of the sciences, and also those who are engaged with community development and advocacy, and the lessons learned from this book will also provide for continued personal growth. This book also brings to its readers the concepts and skills necessary to achieve a better understanding of how your specific project goals can be best achieved in the current rapidly integrating disciplines of the environmental sciences, management, and policy by providing explicit examples that illustrate a variety of encountered situations and solutions, across the ever-changing global human landscape. Each of the global stories offered in this book provides clear depth and a variety of author's/practitioner's perspectives from the Asia-Pacific Region, to the Americas, to Africa, to the Middle East Region.

The discerning reader will definitely notice that there are many opportunities to combine the conceptual and operational linkages described within this book, which outlines the integration of societal needs and circumstances with very specific parameters of scientific endeavors. The imaginative reader will also find several opportunities to make connections between and among their own specific natural resource management or other environmental issues and goals, and utilize the specific examples, techniques, and approaches provided among the eight chapters of the book to build a repertoire of new methods for more effectively addressing the current tasks before them.

In some ways, the stories in this book are not exceptional, in that they are no more special than the stories we all keep in our memories, yet the stories shared in this book are also quite special and rich in lessons. As are all stories about the problems that arise in the lives of real people, the stories in this book are compelling because they describe our common goals as a species, allowing us to, perhaps, improve the lives of others across this complex and varied planet. No doubt you have a set of unique stories to recall too, perhaps related to some of the aspects outlined in this book, i.e., either a societal problem needing a solution or a science approach aimed at solving that problem, or both. The following chapters are intended to help you place those stories in a context that will allow you to see more clearly the value of the challenges and successes you may have encountered in your life/work, and provide you with examples of how similar integrative work has resulted in transformational change for communities and science applications. We, the authors of this book, have selected particular stories that specifically demonstrate how the interactions among collaborators developed, how the collaborative energies grew among participants, and specifically how the projects came to fruition. We ask that you as the reader think of the commonalities among all of our collective stories and reflect upon how the examples in this book may enhance the steps you took, or are planning to take, to complete your particular collaborative environmental work within your communities of interest.

Note that not every one of the projects within the following chapters went completely smoothly, and without impediments. In fact, those impediments that did arise in each of the stories in the following chapters is a key lesson unto itself, which I would recommend to you as the reader to focus upon, i.e., take a close look at these "failures" (and see Section 8.3) because it is those so-called project failures that typically lead teams to struggles, then to a point of reassessment of the purpose and scope of the work involved, and consequently to then build strength within the team, ultimately making for a better outcome in the long run. In each of the following stories of collaboration and transformation, the struggles are what built the unique strengths of the work; you will see how, without such so-called failures, there are rarely the robust outcomes that one might hope for with any societally relevant solution. Without this enlightened view of embracing "failures," and consequently a more collaborative view of meaningful utilization of the knowledge of the work team, there is a greater chance for the work team to be diminished in purpose, specifically as a result of missed opportunities for discovery, which ultimately can diminish the likelihood, the impact, or the completion of potentially transformational work.

By way of sharing stories, such as those you find in this book, we can increase the consciousness of anyone reading them, and in the case of this book, our intent is to better describe the roles of scientists in addressing specific societal issues; by making these stories more accessible to everyone, regardless of their particular background, education, or perspectives on life, we tend to broaden future discussions that integrate science and everyday needs of people. To this end, we focus on an aspect of the environmental sciences that is sometimes ignored, i.e., the human factors, making this book a first of its kind, in that it is focused primarily on the needs of people, and then secondarily on the core work that societies are using to address their needs. The stories in this book aim at the specific roles that a wide array of environmental scientists featured throughout the chapters (e.g., within the disciplines of geology, ecology, natural resources, biology, agricultural science, food science, and hydrology) play in a variety of societies around the world. One might suppose that the book's focus is on decision makers, which is a commonly-referred-to audience of technological and research outcomes in the environmental sciences. However, this supposition would be false, in that this book is not concerned with any one particular audience of science information, but rather it is focused on society as a whole, and how society utilizes all forms of information, in all of its varied forms of knowledge or "ways of knowing." As such, this book takes a straightforward humanistic approach to understanding how environmental science work is accomplished, an approach that I (Lopez) and others have employed throughout our careers—an approach of straightforward and unassuming dialog with everyday people, focused on listening. During such dialogs with people, about their needs, discoveries are made on all sides of the dialog. Such discoveries are the magic that provides a special connection between the, occasionally and necessarily, dispassionate and objective world of science and the, frequently, warm and caring world of people, their families, and the lives of the groups they associate with, such as people of a specific locale or town or village, members of a specific ethnic group or interest group, or an assemblage of colleagues that are focused on similar values and goals. It should also be noted that there is often a special connection that occurs between people and their place(s), meaning the land(s) that they feel tied to, or otherwise close to. This "sense of place" is very important to most people around the world, based upon their specific ties to the land, and is a common theme and powerful driver of all of the stories within this book.

Examples of dialogs that center around the variety of key connections that occur between and among people, and place(s), are seen in each of the chapters to follow, each from a very distinctive set of societal, cultural, institutional, and professional perspectives. Since this book is designed to be a general text for the use of many, including scientists, community members, and practitioners, each of the dialogs encountered within the chapters of this book should strike you in a familiar way, hopefully stimulating different recollections and other aspects of your interactions with environmental science topics, members of communities, and work teams. As you read the following stories, please take plenty of time to note the variety and scope of each of the case studies provided, and then ask yourself how the societal, cultural, or other differences among the participants in the specific case study furthered the goals of the projects described. You will likely find that the answer that you come up with is similar to some of the answers arrived at among several of the chapters of the book. After all, we all share very similar needs for scientific information, regardless of the people or communities involved in the dialogs. Whether the concerns of society are in central Iraq, Southeast Asia, a small island in the western Pacific, or within the rolling hills of the American Midwest, one common thread you will notice as you make your way through these stories is the need we all have for improved information about our world and our common desire for relevancy of the work we do as scientists, practitioners, and community members.

1.2 A Quick Guide to Your Global Journey

This book is subdivided into eight chapters, each with a specific intent to fulfill an explicit piece of the larger story of collaboration and transformation as a whole, and intended to build upon its previous chapters. Each chapter is uniquely presented by the author(s), who are themselves members of the communities in which they do their work. A brief summary of each of the chapters, a miniature journey around the world, follows:

Chapter 1 prepares us for the concepts of more integrative approaches to solving societally based problems using a variety of knowledge, including traditional knowledge as well as the environmental sciences. It is intended to prime the reader for the concepts to come in subsequent stories from across the globe, with a specific example of collaboration and transformation from Hawai'i, a land of diverse interests, histories, and people. This story epitomizes the concepts of embracing "different ways of knowing" in order to specifically address a contemporary societal imperative, a sudden and extensive loss of a culturally and ecologically significant tree species.

Chapter 2 transports us to the other side of the globe, to the cradle of civilization, Mesopotamia—modern-day Iraq. Iraq is a resource-rich country, but years of war and conflict have hindered the development of its agricultural, water resources, mining, energy, tourism, health care, and educational sectors. Since a fairly recent decrease in active war, the increases seen in oil production have not kept up with falling oil prices, compounding economic woes for the people and government of Iraq, resulting in large budget deficits, economic recession, and other associated societal stresses. Postwar, food security, and access to safe water remain key environmental problems for a large part of the population of Iraq. The government is calling for foreign investment into the renewable energy and mining sectors, as the country faces a barrage of overwhelming environmental problems, including severe pollution from two decades of war, international sanctions, and mismanagement by the Saddam Hussein regime. The United Nations declared that "many environmental problems in Iraq are so alarming that an immediate assessment and a cleanup plan are needed urgently" (UNEP 2005), creating a societal imperative that is very compelling, yet so devastating that one may not know where to begin. Despite this rather dire situation for the people of Iraq, and in the face of tremendously expensive and complex environmental remediation challenges, a number of breakthroughs have occurred over the past several years. Through persistence and collaborative approaches that involve both national experts and expatriates, many in Iraq and internationally are now focused on returning traditional landscapes to what they once were in Iraq, both ecologically and societally, in an attempt to restore the land and lifestyles that existed prior to the tumultuous two decades of war and strife within Iraq. One of several breakthroughs came in July 2016, when the marshes of southern Iraq were granted World Heritage status by the United Nations Educational, Scientific, and Cultural Organization after several years of collaborative work by the people of Iraq. This chapter chronicles this success and the future challenges of this historically rich land, and its people, thereby inspiring even the most pessimistic among us who may also be faced with challenging environmental and societal issues within our own community.

Chapter 3 transports us back to the other side of the planet from Iraq, taking us on an intentional detour to delve deeply into the minds of people and what motivates them to do extraordinary things to bring stewardship to their changing landscape. Chapter 3 is a fascinating look at the hearts and minds of the people of the western Ozarks, quite literally "Middle America," over a 25-year period from 1993 to 2018. Chapter 3 reflects and celebrates the story of the Upper White River Basin in the Ozark Plateau of the states of Arkansas and Missouri. The people of the Ozarks tell much of the story, a unique cultural perspective on water, and their efforts to protect both the watershed and their hopes and dreams for a better life. The karst geology and somewhat isolated geography of this heavily wooded and mountainous region is briefly explored before turning to a series of graciously given reflections on the successes and celebrations surrounding the water that their lives center around, and its profound impacts it has upon their lives. Chapter 3 also reflects on a few deep lessons learned by the people of the Ozarks and their particular cultural wisdom that resulted in an organically-formed and dedicated cadre of people, known and unknown, playing central roles in improving and protecting the health and vitality of the Upper White River Basin.

After learning of the challenges and the breakthroughs in Hawai'i, Iraq, and the Ozarks, Chapter 4 subsequently takes us to the far western reaches of the Pacific, to a group of islands in the State of Yap, within the Federated States of Micronesia. Because islands are ecosystems that are thoroughly understood, at a human scale, by those who live there as a unit of finite resources, island societies are often even more keenly focused and integrated around their environment, compared with societies on large land masses. Accordingly, the scientists residing on islands are not just part of a community of science peers but also a part of the island community itself. Chapter 4 addresses this interwoven network of science and society by providing a detailed account of science that is, thus, intermingled with the development of environmental efforts necessitated by the islands' increased commerce with the wider world and the global challenges that emanate from that set of realities. The story in Chapter 4 follows a sequence of fusions of social and scientific responses to general challenges of economics, climate change, and efforts to safeguard biodiversity, to a more specific focus on integrating traditional knowledge and modern scientific technology, all for the societal imperative of maintaining food security and the multitude of ecosystem services that are understood generally as "biodiversity" on the islands of Mainland Yap.

Chapter 5 takes us to a new realm of discovery, by dealing with a global community of practice and their food security, a primary need for all humans. The unique applicability of this story is to be especially noted, in that it does not solely focus on a specific geographic location, but rather utilizes a seamless integration of biochemical analytical techniques in the agriculture sciences and food sciences, with the laws of God, for Muslims worldwide. This particular science is providing information to a "community of practice," serving their contemporary societal need to know, with high confidence, the quality of what they eat and drink, since rarely do we now produce our own consumables, nor do we understand with certainty the food supply chain. Because Islam requires a specific code of conduct and rules for the food supply, the work in Chapter 5 describes how a seamless and highly sophisticated approach is being taken in Malaysia to integrate science with religious beliefs, demonstrating how the global community of 1.8 billion Muslims are served by the environmental sciences, reinforcing trust in the integrity of the food supply, through knowledge and research and development coordination.

Chapter 6 deals with a fundamental driver for most contemporary societies around the world, i.e., the development of new energy sources, and it delves into the environmental impacts that these developments may have on society and a variety of perspectives about those impacts. Chapter 6 returns us to North America for this part of our journey, focusing us on an area of the country that has some unique energy-driven issues, overseen by the United States Environmental Protection Agency (USEPA) and their targeted efforts to address both the science of, and the stakeholder participation in, ecosystem conservation as it relates to this energy sector development issue. This chapter explores a specific technique for engaging impacted communities in the potential landscape changes of the American Midwest that would be created by increased production of biofuels, through a series of stakeholder meetings that focus on "future landscape scenarios" predicted by Midwestern biofuels production. Unique data analyses and modeling efforts were initiated, and several small groups of community members were established such that participants discussed and answered a set of questions about what they perceived the future would hold. For example, what are your views on the quality of life, what do you value from living in the Midwest, what trends for the future do you find most hopeful or most worrisome, and how would you like agriculture to be different? The partnership between the USEPA and the public ensued, and the outcomes of this approach are instructive and provide an approach that may apply to other communities or societal structures.

Chapter 7 once again whisks us across the globe to several areas with a common issue, in Southeast Asia, focusing us on an important and final global story of community involvement as a way to demonstrate true collaborative approaches for solving environmental issues, and indeed transforming societies for the 21st century's issues surrounding sea-level rise. Chapter 7 demonstrates, through comparative case studies in Việt Nam and Cambodia, nuanced differences in the sociopolitical needs of communities in Southeast Asia, using similar scientific approaches. Communities in these areas generally benefit from intact and healthy mangrove forested ecosystems, which directly affect the health and well-being of the people of this vast region, specifically in terms of secure food production, fiber production, fuel production, protection from storms, other economic values such as tourism, and climate change mitigation. The authors outline their contributions to, and the overall global perspectives on, community involvement in managing mangrove forests.

Chapter 8 succinctly synthesizes the large and varied amount of knowledge contained within the book's compilation of work by reflecting on the topics of the entire book and synthesizing the key themes. During our vast journey across the globe in this book, several key "collaboration and transformation" themes arise, and highlighting these key themes at a "high altitude" can be instructive, for taking the broader perspective on the "global human landscape" allows us to draw lessons that can be applied to a number of other types of situations, project, and geographies. Implications for the future of the societal dimensions of environmental sciences are highlighted in this final chapter, which seeks to round out the totality of new knowledge conveyed in the full complement of chapters of the book. As we finish up in the last chapter by establishing some linkages among the topics of the book, you as the reader may care to return to the individual chapter materials and explore them again, specifically to compare to your personal work, or to contrast among the other global stories in this book.

1.3 An Impetus for This Journey into the Societal Dimensions: A Story from Hawai'i

Up to this point, you may have been reading the words I have written in a passive and agreeable way, accepting the conventional understanding that most of us have about the importance of collaboration, which is harmless enough and a very natural reaction to such a reasonable topic, i.e., collaboration. Who can argue with collaboration, after all? Despite this general acceptance of the concept of collaboration, the dynamic that occurs within today's successful societally relevant environmental science projects requires some additional energy and intention to further the concepts of collaboration because it is no longer sufficient to merely pay lip service to collaboration; one must really try hard to ensure there is a quality connection between the many participants in environmental science projects, whether as a consultant, a government agency professional, an academic, or other decision maker. To "lean into" this concept of collaboration in the environmental sciences, we must initially consider that it is not a simple task to communicate among such diverse participants (i.e., of varying cultural, educational, gender, and other perspectives), which tend to be involved in all environmental project teams and who possess a variety of different "ways of knowing" the world.

To break the ice a bit in preparation for the upcoming Chapters 2–7, ahead, in this chapter we will kick off the conversation about communication between a variety of people with differing ways of knowing by describing a specific journey we (the authors of Chapter 1) had, which improved our awareness of the critical importance of community collaboration. This awareness has also slowly shaped my (Lopez) increasingly intentional and focused approaches to science, during my career, which culminated in the past several years of work at/with the United States Forest Service's research and development group in Hilo, Hawai'i. To follow is a brief reflection of a journey toward integrating societal needs and the environmental sciences, a consciousness-building period of my career, which was the impetus for sharing a variety of other stories of collaboration and transformation, throughout the remainder of this book. The experience that we will share in this chapter is from Hawai'i, during which time we had the pleasure of working with many and diverse communities within the Asia-Pacific Region. Part of this story involves my (Lopez) personal reflections as director of the research institute in Hilo, Hawai'i (those reflections I will refer to in the first person), during which time I had occasion to focus on intentionally serving the traditional and indigenous communities of the seven U.S.-affiliated political entities, which the research institute is charged with serving, in the following locations: the State of Hawai'i, the Territory of Guam, the Territory of American Samoa, the Commonwealth of the Northern Mariana Islands, the Republic of the Marshall Islands, the Federated States of Micronesia, and the Republic of Palau.

The following example from Hawai'i is emblematic of many other stories in the Pacific, which was an important point in my awakening about the critical importance of collaborative work to affect the change in societies where there are environmental problems that need a solution. The experiences in Hawai'i provided me with a special understanding of the mechanisms of the interrelationships between societal problems (not the euphemistic "challenges," but rather, true problems!), the people that have these problems, the scientific approaches one might choose to address the problems, and the special communication necessary to integrate the solutions based upon the needs of the community. My enhanced consciousness that has evolved over time about this subject, including the pivotal time in The Pacific, brought forward the clear importance of the interrelationships in the example in Section 1.4, which provided me a focused opportunity to intentionally utilize mechanisms of information sharing, respect for multiple ways of knowing the problem and potential solutions, and, importantly, patience and dedication to the problem, shared by many. The example I am talking about is a very challenging societally important environmental problem, an infestation of a deadly fungus that has been killing (by way of the plant disease known as "Rapid 'Ōhi'a Death") a vast portion of a culturally, societally, and ecologically significant tree species in Hawai'i, since approximately 2013.

1.4 A Societal Imperative in Hawai'i: Dieoff of 'Ōhi'a Lehua

Societal imperatives come from true problems, not the euphemistic "challenges" that we often ascribe to issues emanating from various sectors of society, but rather the true problems that are of a magnitude that make people not only emotional but also powerfully focused on action, and at times even becoming a bit distressed about the issue at hand. It is this type of problem, the die-off of a tree known in Hawai'i as 'Ōhi'a Lehua (Figure 1.1), or commonly just 'Ōhi'a, that I encountered during my time as director of the Institute of Pacific Islands Forestry (hereafter, The Institute) in the 2013–2018 timeframe, a true societal imperative.

1.4.1 'Ōhi'a's Imprint on Culture and Ecology

'Ōhi'a Lehua, *Metrosideros polymorpha*, is one of the world's best examples of a "foundation species." It is a tree species essential to native forest ecosystem functions, and it harbors the rich endemic biodiversity of Hawai'i's forests. Consequently, 'ōhi'a has long-held enormous cultural significance for native Hawaiians. Kumu (meaning "teacher" in the Hawaiian language) Sam Gon described this unique role for 'ōhi'a as follows:

> ...the dominant tree of the majority of forest communities in Hawai'i, it is no surprise that 'ōhi'a lehua and other *Metrosideros* species bear great Hawaiian cultural significance. The tree itself is considered

FIGURE 1.1
The culturally, societally, and ecologically significant tree species in Hawai'i, 'Ōhi'a Lehua (*M. polymorpha*).

the kinolau (physical manifestation) of Kū, one of the four principal Hawaiian deities, though 'ōhi'a may also serve as one of the kinolau of several other Hawaiian deities such as Laka (patron deity of hula) and Pele (goddess of volcanism). Its wood is incorporated into the two most sacred structures of the heiau (temples) of governance: the ki'i akua (god figures) and the lele (offering platform). The flowers of 'ōhi'a, called lehua, are symbolic of the fires of Pele, the blood of warfare, and hard-won, long-trained expertise. The role that lehua blossoms play as the mainstay of Hawaiian honeycreepers and other nectarivorous birds was not ignored; the attraction of birds to lehua is used as a metaphor of courtship, popularity, and high regard. The bog form of lehua has a spe-cial name, lehuamakanoe, which denotes its mist-shrouded habitat and confers a sacredness afforded only to deities. Material and other cultural uses incorporate the wood of 'ōhi'a in houses, tools, weapons, religious structures, food preparation tools and containers, the foliage and flowers in lei, and the tender parts of flowers, foliage and roots in herbal medi-cine. As a prevalent and easily recognized presence in ecosystems from sea level to tree-line, numerous 'ōlelo noeau (wise and poetical sayings) refer to 'ōhi'a as a positive symbol of strength, sanctity, and beauty. All of these primary cultural underpinnings should not be ignored when considering the value of 'ōhi'a as an element of essential biocultural conservation value (2012).

The interwoven character of 'ōhi'a is captured in the succinct, rich Hawaiian mantra, "I ola 'oe, i ola mākou nei" (translated as "my life is interconnected to yours, your life is interconnected to ours"). Additionally, Kūka'ōhi'alaka, one of many deities connected to the natural world in the Hawaiian culture, specifically represents the multiple life cycles within the native forest and is portrayed in the male energy of hula dancers. More explicitly, he represents the physical process of evapotranspiration, the living breathing forest, and most importantly represents 'ohi wai—the mist collector that captures water from the air to feed the aquifers. To this day, we can see the significance of the hula adornments connecting humans to the 'Ōhi'a Lehua (i.e., lehua here refers to the flower blossom). These adornments reveal the process of becom-ing connected to the forest and the elements represented therein. The danc-ers are able to transcend space and time, for the dancer wearing 'Ōhi'a Lehua becomes the associated elements in nature that we see to this very day.

For most in Hawai'i, the cultural importance and the ecological impor-tance of 'Ōhi'a Lehua go hand in hand, and indeed the cultural and ecological elements of the land ('āina) are not separate. Yet, for the sake of discussion, we are describing the ecological importance of 'ōhi'a here distinctly so as to call out the specific biophysical reasons that the loss of the tree is not desir-able for society at large in Hawai'i. 'Ōhi'a, a member of the Myrtaceae family (*M. polymorpha* Gaudich.), is the most abundant and widespread native tree in the Hawaiian Islands, occupying approximately 80% of all native forest statewide. Although Hawai'i has lost roughly half of its original native forest, about 350,000 ha of 'ōhi'a-dominated forest remains, of which approximately

250,000 ha occurs on Hawai'i Island and 100,000 ha occurs on the other main Hawaiian Islands. 'Ōhi'a is also among the best existing examples of a "foundation species" (Ellison et al. 2005), meaning it is essential to native forest ecosystem function and to the native biodiversity that Hawai'i's forests foster. Some have also preferred the term "keystone species" to describe ōhi'a's fundamental ecological role in the forests of Hawai'i.

'Ōhi'a has been evolving in the Hawaiian Islands for nearly four million years and has become genetically adapted to diverse local environments (Percy et al. 2008). Originating in the Marquesas Islands from an ancestral species of *Metrosideros* that colonized new lava flows in New Zealand, 'ōhi'a was apparently able to outcompete most other plant species in precontact Hawai'i and came to form the habitat conditions for the evolution of a large fraction of Hawai'i's endemic flora and fauna. Although some other important plant groups underwent complex adaptive radiation, vigorous gene flow apparently kept Hawai'i's 'ōhi'a from radiating into multiple species.

'Ōhi'a is a remarkable generalist, with the ability to dominate a broad range of sites in Hawai'i—between sea level and 2,500 m in elevation, and from very wet (up to approximately 6,000 mm/year [236 in./year]) to relatively dry (approximately 500 mm/year [20 in./year]) environs. It varies from an early colonizer species on Big Island lava flows to long-term dominance on four-million-year-old soils on Kaua'i. It has been evolving in place for a long time and has adapted genetically to local environments, but specialization is limited by within-species gene flow (Vitousek 2004).

Although multiple species and varieties have been delineated and named by various botanists, Percy et al. (2008) concluded that there is a single highly polymorphic species of *Metrosideros* in Hawai'i: *M. polymorpha*, based on literature and analysis of chloroplast DNA. However, after additional analysis, the same research group (Harbaugh et al. 2009) suggested that, based on nuclear DNA microsatellite loci, there are possibly five species-level genetic groups in the Hawai'i *Metrosideros* complex: *M. polymorpha*, *M. dieteri* and *macrophylla*, *M. rugosa*, *M. tremuloides*, and *M. macropus*.

Gruner (2004) recognized the preeminent ecological role of 'ōhi'a, particularly regarding Hawai'i's native fauna; it can be argued that *M. polymorpha* is the backbone of Hawaiian forests and one of the most important resources for the long-term stability of ecosystems and watersheds among the islands. *Metrosideros* is an important, year-round nectar resource for native bees, moths, thrips, and other insects, and for native nectarivorous birds. Numerous insect species use 'Ōhi'a Lehua as a resource for either food or habitat space, and it may have the largest associated fauna of any native plant.

Forests dominated by 'ōhi'a are home to at least 22 extant species of forest birds, the Hawaiian hoary bat, and many of Hawai'i's remaining native plants and invertebrates. Endemic Hawaiian honeycreepers, including eight on the endangered species list, are dependent on these forests for essential habitat, as they have adapted to feed and nest in 'ōhi'a trees (Pratt et al. 2009).

The widespread or total loss of 'ōhi'a would spell disaster to forest birds, certainly ensure extinction of multiple species directly dependent on 'ōhi'a, if not most species, given the foundational role that 'ōhi'a plays in the forests these birds depend on. Three nectarivorous bird species ('apapane, 'i'iwi, and 'ākohekohe) heavily depend on 'ōhi'a nectar blossoms; even though they visit flowers from other plant species, there is no substitute for the volume, geographic spread, and year-round source of nectar 'ōhi'a provides.

Loss of 'ōhi'a forest areas would also impact non-nectarivorous bird species. Most species preferentially nest in 'ōhi'a, even koa specialists such as the 'akiapōlā'au. Insectivore species such as Hawai'i 'ākepa and 'akeke'e are specialists on 'ōhi'a leaf and flower buds, and most insectivorous birds preferentially forage in 'ōhi'a trees. Even bird species that may not be entirely dependent on 'ōhi'a for food resources (e.g., 'elepaio and 'ōma'o) would likely be greatly affected by the substantial changes in native forest structure that would ensue with widespread 'ōhi'a mortality. In short, the loss of 'ōhi'a from Hawai'i's forests would negatively impact native birds, with the amount of harm depending on the percent loss of 'ōhi'a trees. The complete loss of 'ōhi'a on all the main Hawaiian Islands would likely lead to the extinction of most of the extant 19 native Hawaiian bird species dependent on 'ōhi'a-dominated forests.

Nearly 500 different endemic Hawaiian species of arthropods (i.e., insects) have been collected from 'ōhi'a canopies (Gruner 2004), further demonstrating the rich biodiversity that 'ōhi'a forest fosters. In addition, several of the largest endemic insect groups that associate with plants (e.g., leafhoppers, planthoppers, and psyllids) contain species that rely exclusively on 'ōhi'a and no other plant species (Zimmerman 1948).

'Ōhi'a thus provides the foundational and functional framework for Hawai'i's forested terrestrial ecosystems and the diversity of the native fauna and flora; roughly 90% of which is endemic, occurring nowhere else in the world. Although perhaps 10% of the plant species of the archipelago have been lost to extinction and 30% of the remaining species are endangered, extensive and relatively intact tracts of native-dominated communities survive, primarily at higher elevations.

Not surprisingly for a single species with a very broad ecological distribution, 'ōhi'a exhibits substantial genetic variability (Aradhya et al. 1993; James et al. 2004; Crawford et al. 2008; Percy et al. 2008; Harbaugh et al. 2009; Izuno et al. 2016), phenotypic plasticity (Cordell et al. 1998; Cornwell et al. 2007), and combinations of the two phenomena (Cordell et al. 2000, 2001; Fisher et al. 2007). Phenotypic plasticity refers to the important phenomenon involving single genotypes that can produce different phenotypes (i.e., observable characteristics) in different environments. 'Ōhi'a is well adapted to the large climatic intra-annual variability, particularly with regard to rainfall that is so characteristic of Hawai'i (e.g., Loope and Giambelluca 1998). Scientists hope that the expressed and acknowledged genetic variation and phenotypic plasticity of 'ōhi'a may provide for sufficient levels of resistance to Rapid 'Ōhi'a Death (ROD) in certain 'ōhi'a populations. Research is under way to discover whether or not such resistance is present, but this work needs to be expanded further, to adequately address this possibility.

1.4.2 First Glimpses, Uncertainty, and Awareness

In 2009–2010, residents of the Puna District on Hawai'i Island first began reporting the abrupt demise of 'ōhi'a trees on their properties and adjacent forest stands. On a previously healthy-looking tree, all of the leaves of a major limb or the whole tree would turn yellow, then brown. Within a few weeks, the tree would appear dead. The spread of dying trees was initially slow but accelerated over time. By 2012, 10% or more of 'ōhi'a trees of all ages had died in a patchwork of approximately 1,600 ha (4,000 acres) across the Puna District (Mortenson et al. 2016). By 2014, the affected area had increased to approximately 6,400 ha (16,000 acres), and the problem began receiving major attention and research. Initial sampling of roots of symptomatic trees revealed nothing unusual, but in 2014 the University of Hawai'i's College of Tropical Agriculture and Human Resources (UH-CTAHR) isolated a fungus from wood samples of a symptomatic tree that was identified as belonging to the genus *Ceratocystis*, a genus that includes species that are pathogenic to a wide variety of tree species. By late 2014, pathogenicity tests conducted by Dr. Lisa Keith and colleagues in the United States Department of Agriculture-Agricultural Research Service (USDA-ARS), Pacific Basin Agricultural Research Center, had determined that the agent causing the disease is allied with the vascular wilt fungus *Ceratocystis fimbriata* (Keith et al. 2015). Although several strains/genotypes of *C. fimbriata* have been present in Hawai'i for decades that affect sweet potato, taro, and *Syngonium* (a common ornamental plant in the same family as taro, in the Araceae family), this newly identified pathogen appears to be either a new strain of species that causes the current dieback phenomenon known as "Rapid 'Ōhi'a Death," "ROD" for short. Symptoms of the disease ensue as the fungus invades and spreads within the sapwood of affected 'ōhi'a trees, essentially prohibiting the transport of water and sugars through their vascular systems.

It is not yet entirely clear how the ROD pathogen spreads—particularly how it has spread so quickly. Current hypotheses being tested are that it is spread long distances by human activity (e.g., infected wood or other plant and soil matter, equipment, tools, and shoes) and by wind as spores imbedded in tiny bits of insect frass, i.e., the fine powdery refuse and excrement produced by wood-boring beetles and expunged from miniscule tunnels created within dead or dying trees infected by the fungus. However, it is likely that a confluence of factors may contribute to disease spread. These factors may play more important roles in some areas than in others. Tree wounds likely increase susceptibility to infection by allowing entry of the pathogen, and episodic strong windstorms may be important events that both move infective fungal spores embedded in frass/boring dust and create wounds in trees that allow ingress of such spores inside individual trees. Beetles that bore into 'ōhi'a wood may also carry long-lived spores either in their gut or in their exoskeleton that contribute to the infection of at least some trees. Again, transport of durable spores by humans on tools, vehicles, etc. is a major concern (e.g., locally to Kona, Kohala, and Ka'ū, and potentially to other islands); much care is required to avoid this. Feral pigs and other ungulates are capable of moving the pathogen short to medium distances, but no evidence is available to confirm this possibility.

Trees within a given stand where the disease is present seem to die in a haphazard pattern; the disease does not radiate out uniformly from already infected or dead trees. Initial published data from 'ōhi'a-dominated forests where the disease was found present indicated that mortality was fairly evenly distributed across size classes (i.e., included small to large individuals) and that the average annual mortality rate of 'ōhi'a individuals was about 25% (Mortenson et al. 2016), a rate markedly higher than those documented for most other tree diseases elsewhere in the world. In wet lowland forests of the Puna District, on Hawai'i Island, up to 90% or more of the 'ōhi'a trees in a stand have succumbed to the disease within a span of just a few years. Invasive strawberry guava (*Psidium cattleianum*), *Melastoma* spp., and Koster's curse (*Clidemia hirta*) are in many cases poised to attain dominance following the demise of some areas of 'ōhi'a stands in these areas. Other areas of dead 'ōhi'a in the Puna District have been completely replaced by freshly created lava, the result of the 2018 outflow from a newly formed vent lava from Kilauea, near the Leilani Estates area. Prior to this downward adjustment of dead-standing 'ōhi'a forest areas, augmented by lava flows in the Puna District, and as of February 2016, areas with 'ōhi'a mortality, much if not most believed attributable to ROD, totaled approximately 15,000 ha (38,000 acres), across the Puna, Hilo, Kona, and Ka'ū Districts, with the largest known infestations in the Puna–Hilo–Volcano area. By September 2016, the total area of 'ōhi'a mortality had increased to approximately 20,000 ha (50,000 acres). By the spring of 2018, the total affected area on Hawai'i Island had increased to 135,000 acres (Figure 1.2). That increase is partly due to increased mortality and partly due to more total area having been surveyed. It is now recognized that, in the absence of effective management approaches, the pathogen may have the potential to eventually kill many if not most 'ōhi'a trees on Hawai'i Island and threaten the persistence of 'ōhi'a forests statewide. Currently, however, approximately 85% of 'ōhi'a forests on Hawai'i Island remain healthy and free of ROD, and ROD has only been found in what is thought to be a less virulent form in a small area of Kaua'i Island—all of the other Hawaiian Islands remain free of the disease, as of this writing. There is strong determination and growing capacity by organizations on other islands to quickly detect ROD when it first arrives and to eradicate it. Experimentation is under way on Hawai'i Island to refine the methods relevant to those intentions and to eradicate new localized outbreaks of the disease before they become extensive infestations.

FIGURE 1.2
'Ōhi'a forest on Hawai'i Island seen from low-flying aircraft in 2008 (top inset) and the same area after infestation with fungus that causes "Rapid 'Ōhi'a Death" in 2012 (bottom inset). (Aerial photographs: US Geological Survey.)

Hawai'i Department of Agriculture (HDOA) has taken responsibility for enforcement to prevent transport of ROD to non-infested islands. The Hawai'i Board of Agriculture (on August 25, 2015) passed an interim rule restricting the movement of 'ōhi'a from Hawai'i Island. The interim rule was finalized in November 2016. Agencies and the public are asked to follow many precautions such as not moving 'ōhi'a wood, cleaning tools, to pressure wash contaminated vehicles, and to clean clothing and boots (www.rapidohiadeath.org; checked September 15, 2018).

Much of what we think we know about ROD involves "best guess" hypotheses, based on common sense observations and educated experience from elsewhere. These hypotheses need to be confirmed and refined or refuted and replaced through research. Meanwhile, there is consensus that immediate experimental management actions are warranted, accompanied by research, with a promising opportunity to prevent the worst from happening with ROD.

Continued environmental science research is crucial and under way on the epidemiology of this phenomenon, to find more rapid and accurate means of detection in wood and soil (through DNA analysis), vectors for spread, environmental constraints, monitoring by remote sensing and plots, management, and treatment options. Natural host resistance or tolerance is one of the promising approaches being explored and tested. Prospects for developing resistance in 'ōhi'a will be pursued if necessary, with the aim of eventually restoring pathogen-resistant 'ōhi'a forests to the extent feasible. Given that a full operational program for restoration of 'ōhi'a would take decades (and could even fail), it may be important to explore and test the feasibility and desirability of using carefully selected non-native tree species to compensate as much as possible for a disastrous loss of 'ōhi'a, with the aim of maintaining watershed functions and providing a suitable habitat for endemic biota. Additional means of protecting the genetic variability of 'ōhi'a is through the 'ōhi'a seedbank project, led by the University of Hawai'i Lyon Arboretum and supported in East Hawai'i Island by a combination of community volunteers and staff at the U.S. Forest Service's Pacific Southwest Research Station, Institute of Pacific Islands Forestry in Hilo.

Although *M. polymorpha* is likely the only "foundation species" in the genus, there are more than 45 species of *Metrosideros* outside Hawai'i—primarily in New Caledonia (21 species) and New Zealand (12 species)—but also on many Pacific Islands and in South Africa. Arguably, protection of these other *Metrosideros* species will depend on what we learn and how successfully the ROD outbreak in Hawai'i is contained, or otherwise resolved. Partnering with other countries regarding *Metrosideros* could help Hawai'i develop and execute a more effective ROD control program for the benefit of Hawai'i and other Pacific nations. New Zealand has recently implemented a quarantine to prevent the arrival of *C. fimbriata*, with particular attention to the importation of kiwi fruit and *Metrosideros*.

1.4.3 The Impetus Requires People of All Backgrounds, with a Passion

In the summer of 2015, one of the Institute's research scientists (and a coauthor of this chapter) walked into my office, with a very concerned look on his face. Dr. R. "Flint" Hughes was typically very jovial when he came across from his office in the Research Building of the Institute, to my office in the Administrative Offices. On that day, however, he seemed agitated, and so I told him to come to my office, and I asked him, "what's up Flint?"

> I'm really worried about what I have been seeing in the field with 'ōhi'a, and I believe it is going to be a major catastrophe for the trees that remain if we don't do more.

I had been following this topic, among several others, as an emerging environmental concern in Hawai'i since coming to the Institute in 2011, recalling also that there had been an earlier "'ōhi'a wilt" event in limited areas of Hawai'i Island, some years back. Yet, I thought those events were fairly short-term and smaller extent events, and that the current "wilt" was a similar phenomenon. I had incorrectly presumed this was simply a continuation of the longer standing 'ōhi'a wilt issue and was among a long list of natural resource-related invasive species and ecological pathogen issues, which is an ongoing and long-standing battle in Hawai'i. Also, as with most professional research scientists, advocating for particular current topics of focus is customary, and so I challenged Flint to explain to me his rationale for "doing more." Flint explained that his experiences over the past several months suggested that we needed to make ROD research a priority for the Institute, and taking a leadership role on this topic was critical in the short-term, both Island-wide and State-wide. Flint's arguments were compelling, especially when he paused from his usual jovial persona and explained to me, "you know Ric, in 10 years when these trees are dead, I don't want people to ask, 'where was the Forest Service on this one, why were they on watch and still let this happen?'" I paused along with Flint, looked out my office window that has a beautiful view of 14,000-foot-plus (4,267 m) Mauna Kea, and saw the two 'ōhi'a that happened to grow outside the window there: one fully leafed out, beautiful, and growing; and the other withered and possibly dead from ROD (it is never a certainty until the diagnostic tests are done, but always a possibility that a tree is dead because of the fungus). So, from that simple, passionate, and intentional conversation, between scientists, Flint and I made plans to be the conveners for what was to become a critical working group for addressing ROD in Hawai'i, the "ROD Working Group." Quite honestly, many of the science organizations and management agencies did not want to take on the topic as the convener, in that ROD, like most catastrophic pathogens, was thought to likely to go on for years and not result in anything positive—after all, the topic has the word "death" in it. Politically, the agency that may take on the role as convener for addressing an aggressive and devastating environmental pathogen might be seen as somehow responsible for any mishaps, some had explained to me, and others had expressed to me, "why bother, since there is nothing that can be done, either ecologically, nor politically?" warning me, "my advice is to be very careful in how you take on this role as convener." However, from the initial passion of Flint and myself, the two scientists, a large and effective consortium was indeed built through the creation of the ROD Working Group, a consortium that was designed with an initial core group: Dr. Ric Lopez and Dr. Flint Hughes of the Institute; Dr. Lisa Keith of the USDA-ARS; and Dr. J. B. Friday from UH-CTAHR, Extension Office. The initial success, and impetus for the ROD Working Group, was borne out of an immediate and desperate need for the Hawai'i Department of Agriculture (HDOA) to place a quarantine for Hawai'i Island on the inter-island transport of wood or other nursery

products related to 'ōhi'a. This was no minor feat, requiring the full cooperation of the HDOA, and it was an important factor in keeping the disease off
of the other Hawaiian Islands. This initial imperative was achieved, through
close coordination and negotiations with the HDOA, science research community, and others, leading to an early and strong quarantine on Hawai'i
Island. To this day, this early and enlightened effort by the HDOA has led
to there being no other island "hits" for ROD, until a very recent (May 2018)
confirmation of the lesser virulent strain of ROD on Kaua'i, and no other has
confirmed infestations of *Ceratocystis* on other islands to date.

ROD Working Group subgroups expanded quickly after we first convened
the ROD Working Group in August 2015, including the Communications
Action Team (CAT), the ROD Science Team, the Early Detection and Rapid
Response Team, and the other island-specific ROD Working Groups that
have sprung up since the inception of the original ROD Working Group. All
of the subgroups started by passionate conversations between and among
the many participants of the ROD Working Group, i.e., many hundreds of
people who interacted with the ROD Working Group and its members, over
the several years, from community member individuals, to industry and natural resources agency staff, to U.S. Senators and their staffers, and everyone
in between, from all walks of life and professions.

The main point of this portion of the story was to emphasize that a "small
ray of light" (two passionate scientists) can grow into a guiding force for
addressing a very serious threat to environmental sustainability and culturally important traditions, such as was the case in Hawai'i regarding ROD.
Of note too is the transdisciplinary approach (Rosenfield 1992) and remark-
ably democratic style of the ROD Working Group activities (i.e., no person's
voice is excluded from the ROD Working Group, a fundamental, and known,
founding precept of the group, which is often reiterated by the (now) facilitator, Rob Hauff, with the Hawai'i Division of Land and Natural Resources).
Section 1.5 discusses further the details of the societal transformation that
was brought about by this initial, seemingly small, conversation between
research scientist and institute director, and the subsequent excellent work
of the ROD Working Group and the subgroups mentioned above.

1.5 Robust Collaboration as a Mechanism for Change

ROD elicited a collaborative interagency response in both science and
outreach with an outreach effort that was exceptionally diverse and very
effective at both addressing the science information needs and involving
very deep community connections. A true transdisciplinary approach was
used, which allowed partners to use their knowledge, skills, and abilities to
fill any gaps in understanding, from various perspectives, including both

scientific and traditional knowledge. One example of outreach for ROD was the early engagement of UH-CTAHR's Extension Office forester, who created a website focused on the disease in May 2015 (originally called www.ohiawilt. org and eventually changed to www.rapidohiadeath.org); both URLs for the same website are functioning and utilized actively by all members of the public who are concerned with ROD (checked September 15, 2018). The site hosts highly informative videos, frequently asked questions, maps, research/application summaries, and links to research publications. UH-CTAHR also supported the hiring of a new assistant Extension Specialist in Hilo to focus solely on ROD outreach.

In addition to outreach, ROD researchers have individually and in groups given countless presentations, including at annual Hawai'i Conservation Conferences and at the Hawai'i Volcanoes National Park science talks, as well as small community meetings throughout Hawai'i, with frequent updates. The scientists have also given numerous talks and presentations to various legislative groups, including county councils, the statehouse, the mayors' and governor's offices, academic institutions, and other agencies, assuring that most conservation managers, community leaders, elders (kūpuna), scientists, and other decision makers in Hawai'i know the basic information about ROD, and that they are also able to assist informally in outreach. Presentations on ROD and 'ōhi'a are also given regularly by the network of outreach specialists that are contributing time and effort in an ad hoc outreach working group (i.e., the CAT and other associated groups).

The outreach working group recommended branding the disease as ROD to generate a sense of urgency and clarity, instead of the more scientifically accurate description of the disease as a "wilt." The outreach group identified a rough outline of actions and needs including a list of target audiences, key messages, presentations, and outreach events, both on Hawai'i Island and around the state. The group developed, printed, and disseminated brochures and other print and online outreach materials, including the ROD Facebook page launched in October 2015, which is managed by members of the outreach group with consultation from the ROD Science Team (www.facebook.com/rapidohiadeath; checked September 15, 2018). Outreach content was originally designed to simply communicate the latest science, the quarantine, a sense of urgency, and a sense of empowerment and hope, focusing on actions people could take to slow or stop the spread of the disease.

In addition to reaching forest users and lovers, the conservation community, and other audiences such as inter-island travelers with their messages, the outreach working group recognized both the potential to communicate to a larger audience about ROD and the potential threat of accidentally spreading the disease throughout the state through the world's largest hula festival, the Merrie Monarch Festival (www.merriemonarch.com; checked September 15, 2018), which draws thousands of visitors to and from Hawai'i Island from both inside and outside the state each year.

While recognizing 'ōhi'a's cultural importance and the unique opportunity for outreach and engagement presented by the festival, the outreach team advised against telling cultural leaders what they should or shouldn't do regarding ROD. Instead, the team focused on working with community and cultural leaders associated with the festival to share information about the disease in order to enable hula and cultural practitioners to make their own decisions about actions they could take to reduce the risk of accidental spread. For example, Hawaiian community leader (Kumu Hula) Kekuhi Keali'ikanaka'oleohaililani's letter to the hula community (Appendix A) was a way of honoring the traditions of Hawai'i and seeking a way of understanding and addressing the ROD issue in a "different way" than had been attempted, up to that point in time. An excerpt from Kumu's letter is as follows:

Welina ke aloha dearest Hula people of Hawai'i & the world,

Our profound aloha to each and every ~ 'Olohe, Lo'ea and Loiloi Hula, Kumu Hula, Po'o Pua'a, Alaka'i, Paepae, and Haumāna Hula! Aloha pū to the 'ohana, musicians, singers, chanters, lei makers, researchers, dresser people, seamstresses, and hair and makeup specialists who help us recreate and relive every hula, mele, and story in the best way.

My name is Kekuhi Kanae Kanahele Keali'ikanaka'oleohaililani. I am a granddaughter of a most beloved Tūtū and Kumu Hula who taught me, in every moment of my time with her, to BE IN LOVE with, and have a huge ALOHA for all of our 'ohana. She dedicated her time with me to the simple task of making sure I knew how important ALL of our 'ohana are.

When Grampa took us visiting with family around the island Gramma would say, "this is your 'ohana," then sit down to wala'au or talk story. When we visited Hā'ena, Halema'uma'u or Punalu'u she told me, "this is your 'ohana," and then told stories of those places. At the ocean when we fished she would say, "this is your 'ohana," and offer a chant to the water, and the fish on our hooks. At the māla she would say, "this is your 'ohana," and sing a song to the taro. At the forest she would pick liko from the 'Ōhi'a, hold it in her fingers, and say, "Kekuhi, this is your 'ohana," and we would give a chant of thanks.

This memory has become an every moment practice ~ I continue to dedicate my life to dancing, chanting, singing, and teaching how to Aloha ALL of these 'ohana... like many of you do in big and little ways through HULA. I am writing because our hula 'ohana, the tree that is most used in our art form, the tree that is most responsible for making sure that we have water, the tree that is most used in the carving of ki'i, the tree that we can find on almost every landscape on our island, the tree that many of our bird people depend on, the tree that Hopoe and Hi'iaka made lehua lei from, the beloved 'ŌHI'A... the tree that my Gramma introduced to me as 'ohana ~ is being made sick by a fungus, *Ceratocystis fimbriata* (aka Rapid 'Ōhi'a Death or ROD), that lives both in the soil and in the tree. OUR 'Ōhi'a needs our attention, our awareness, and our aloha.

We don't know how this particular strain of fungus got here or how to treat it properly... yet. But, we do know some simple, but effective things that everyone can do to help prevent the spread of the fungus to healthy trees on Hawai'i Island and to other islands.

My friends of the forest from around the island and I are working together to care for and find ways to heal our beloved 'Ōhi'a… the 'Ōhi'a deserve that~anyone who has ever been sick deserves that! Here are some simple preventative actions that we as Kumu, dancers, lei makers, and 'ohana can do:

- If you need to collect 'ōhi'a wood avoid areas known to have the fungus, or areas that look to be sick from the fungus.
- Leave your lei and kūpe'e, that have liko and lehua in it, on Hawai'i Island. When returning your lei and kūpe'e to our forest, return it to the same area that your liko and lehua came from. Or return your lei and kūpe'e to any of the forest areas identified on the attached map. In doing so you are giving your mana to a forest that needs healing.

For the 2016 Merrie Monarch Festival in Hilo, Hawai'i, the Hawai'i Island community is hosting the Pua'ena'ena Ceremony. This fire ceremony will provide a way for people to offer their kinolau, hakina, lei, and kūpe'e with thoughts of full recovery for our 'Ōhi'a to the fire of Ke Ahi O Hi'iaka (see details on invitation announcement).

- Brush clean the dirt from your shoes and any tools you used, and spray everything with 70% rubbing alcohol when entering and leaving the forest (visit the DOFAW [Hawai'i's Division of Forestry and Wildlife] office in Hilo to learn more, and for some supplies).

DOFAW identified ho'iho'i forest reserve locations

The Division of Forestry and Wildlife ho'iho'i zone (in red on map) located three miles from Highway 11 (Kanoelehua) along Stainback Highway to North Kūlani Road. Please ho'iho'i no more than 20 ft in from forest edge.

The Division of Forestry and Wildlife ho'iho'i zone (in red on map) located between mile markers 12 and 16. Safe pull-outs are located at MM 12 on right side of road, and at MM 16 on left and right sides of road. Please ho'iho'i no more than 20 ft in from forest edge.

- Wash your vehicle, its tires and the undersides if you drive off-road. Keep the inside of your vehicle clean of dirt, and spray the floors of your vehicle with 70% rubbing alcohol.
- Notice and report 'Ōhi'a that are dying in any area that you go to pick, especially areas that are not already on the map of affected forests (contact the folks on the brochure).

Aside from all of those things above, as the hula community is well aware, our songs, hula, chants, prayers, and our thoughts of full recovery sent directly to our 'Ōhi'a community are just as powerful!

Kekuhi

Ulu ka 'Ōhi'a..a lau ka wai

Kumu Keali'ikanaka'oleohaililani's letter showed her aloha and concern for 'ōhi'a as well as suggesting ways for the hula community to be actively involved in preventing the spread of ROD. Her letter was well received and largely heeded by Merrie Monarch participants and the larger community. The outreach group worked together with Kumu Keali'ikanaka'oleohaililani to create a positive cultural message: KŪ! KA 'ŌHI'A LAKA, which draws upon the image of the Hawaiian forest deity, Kūka'ōhi'alaka, who is associated with the multiple life cycles of the 'ōhi'a forests, and mature, healthy 'ōhi'a forests. This message aims to empower people to help in the recovery of 'ōhi'a by promoting the idea that thinking thoughts and saying words of health and well-being can be very powerful, in addition to being aware of the challenges 'ōhi'a is facing and being mindful of field equipment, vehicles, and clothing sanitation protocols for reducing the risk of spreading the disease when going among ōhi'a forests.

The Merrie Monarch 2016 competition organizers and judges understood the urgency that ROD represented, and they allowed late substitutions in proposed lei materials for dancers, which has never been permitted before. Competitors were able to replace 'ōhi'a flowers and liko (new 'ōhi'a leaves) with other plants. The near complete absence of the iconic 'Ōhi'a Lehua from the annual craft fair, the royal parade, and each night of the competition was unprecedented. They also allowed the outreach group to set up an outreach and collection booth outside of the stadium each night of the competition to collect lei and plant material and share information about ROD and Kūka'ohi'alaka (Figure 1.3).

FIGURE 1.3
The impacts of community outreach and working closely with the Merrie Monarch leaders were the near complete absence of 'Ōhi'a Lehua from the annual craft fair, the royal parade, and each night of the hula competition. (Photo: ROD Working Group.)

To prevent ʻōhiʻa products from being transported off of Hawaiʻi Island, the outreach group collaborated with the HDOA on signage, inspector presence, and culturally appropriate ways to collect lei and prevent plant material that had the potential of carrying the fungus from making its way to neighboring islands, with information at both of the major airports on Hawaiʻi Island, in Hilo and Kona. The group also worked with (world-renowned) Kumu Hula, Kekuhi Kealiʻikanakaʻoleohaililani, to create a community event and a new ceremony to celebrate our relationship with ʻōhiʻa and provide a way for the hula community to hoʻihoʻi (return) their lei to the environment without fear of spreading ROD. The first of its kind sunrise Puaʻenaʻena Ceremony took place at the County of Hawaiʻi's Hoʻolulu Complex, adjacent to the festival stadium in the morning after the competition. It was created based on the tradition of making offerings at Halemaʻumaʻu Crater on Kīlauea Volcano, and incorporated fire and a ceremonial burning of the lei material that was collected throughout the week (a practice that is not typically part of the hula world). The ashes of the lei, along with the energy, sweat, and hard work of all the dancers and ceremony participants that went into everything during the festival, were then returned to the forest by outreach group participants.

The Nature Conservancy (TNC) of Hawaiʻi produced a 30-second public service announcement (PSA), narrated by respected Hawaiian Studies scholar Puakea Nogelmeier, which was posted online and aired on KFVE, Hawaiʻi, during the broadcast of the Festival. Through Hawaiʻi's Division of Forestry and Wildlife (DOFAW) partnership with New West Broadcasting Corps, which operates five Hawaiʻi Island radio stations, the outreach team wrote and recorded a 1-minute Big Island Minute radio PSA about preventing the spread of ROD and a 30-second spot which aired a total of 179 times from February 17 to 29, 2016. John Replogle of TNC and Kaila Olson of the Hawaiʻi DOFAW also wrote and recorded a 30-second PSA in Hawaiian pidgin, which played at Pacific Media Groups eight Hawaiʻi Island stations. The PSAs aired a total of 408 times from February 29 to April 4, 2016. In addition, radio personalities announced variations of the script on the following stations: KAPA, KHLO, KKBG, and KPVS—Hilo and Kona.

The University of Hawaiʻi Lyon Arboretum developed an informative video in conjunction with a crowdsourcing campaign (www.gofundme.com/ohialove, now at the University of Hawaiʻi Foundation; checked September 15, 2018) for a project to collect and store ʻōhiʻa seeds to protect genetic diversity during the ROD crisis. Lyon Arboretum staff worked with the ROD researchers and outreach group to ensure consistent and appropriate messaging, and the campaign reached their target funding goal in less than a month (finally funded for $50,030, $30 beyond its goal, taking 31 months and 457 donations).

A partial listing of agencies/groups participating in ROD outreach at this time included UH Cooperative Extension (i.e., UH-CTAHR), USDA-ARS, USDA Forest Service, Kupu Hawaiʻi/AmeriCorps, The County of Hawaiʻi

Mayor's Office, The County of Hawai'i Department of Parks & Recreation, Hawai'i Department of Health, Hawai'i Fire Department, TNC, HDOA, DOFAW, Invasive Species Committees, Coordinating Group on Alien Pest Species, a variety of Watershed Partnerships, and numerous community volunteers. Various informative videos, brochures, rack cards, signs, press releases, and radio PSAs were collaboratively developed, printed, and distributed mainly by HDOA, DOFAW, and the ROD outreach team (i.e., the CAT).

Groups/communities being targeted by outreach included (with coordination by UH-CTAHR, DOFAW, and TNC): the wood industry, the nursery industry, builders who use 'ōhi'a poles, hunters, hikers, ecotour operators, hula/cultural practitioners, hale (traditional home and lodge) builders, arborists and tree services, firewood cutters, the Hawai'i Electric Light Company, the Hawai'i Department of Transportation, county public works, major landowners (e.g., Kamehameha schools), small landowners, politicians/decision makers, and the public at large.

Coordination of work was conducted via ad hoc meetings and a monthly working/planning group call, one per week following the monthly ROD Working Group call. Some representative examples of press coverage for the activities involved in this major effort included the following:

- *Outside* online magazine featured an article about ROD: "What's Killing Hawaii's Trees?" by David Ferry, March 10, 2016: www.outsideonline.com/2060691/whats-killing-hawaiis-trees (checked September 15, 2018)
- "Scientists think insect is likely spreading ohia fungus," *Hawaii Tribune-Herald*, March 23, 2016: www.hawaiitribune-herald.com/?s=scientists+think+insect+likely+spreading+ohia+fungus (checked September 15, 2018)
- KITV, March 23, 2016: www.kitv.com/story/31553119/scientists-testing-beetle-as-possible-culprit-spreading-deadly-fungus (checked September 15, 2018)
- "Scenes of Devastation: Chasing Hawaii's Deadly Ohia Fungus," Molly Solomon, Hawaii Public Radio, March 25, 2016: http://hawaiipublicradio.org/post/scenes-devastation-chasing-hawaiis-deadly-ohia-fungus (checked September 15, 2018)

1.6 The Shift from Reactive to Strategic

Despite the very successful work discussed previously, it was still becoming clear to all involved in the ROD issue that the science and management tools currently in existence were not achieving the eradication of ROD from Hawai'i, in a manner necessary to address the speed and breadth of challenges involved. Furthermore, early projections of the potential loss of 'ōhi'a from infection and mortality rates in monitoring plots were sobering. Therefore, the outreach working group worked together to craft their successful efforts into a strategic plan for the next few years of what was understood to be a long and persistent pathogen. The outreach working group worked up a budget for achieving widespread awareness, support, and participation in reducing the chances of the public spreading the pathogen, with the goal of focusing even more on improving the resiliency of native forest habitat and functions.

One of the main organizational goals was to transition the ad hoc outreach working group into a more formalized structure with several full-time staff to aid in messaging and engagement on each island. Some outreach capacity did exist, at the organic and volunteer level (i.e., collateral duty for people with other job duties), but these personnel must continue to provide outreach for their own organizations about a variety of environmental issues, such as other invasive species, and therefore could only assist part time on ROD outreach in their communities.

The second goal for the strategic plan was to simply and effectively activate and empower people. While the situation surrounding ROD is a very serious one, public outreach and engagement was thought to best focus its messages on what can and should be done to help our forests, not the idea that there is impending doom surrounding the fate of Hawai'i's forests. Similar campaigns have shown that the best way to engage the necessary level of awareness and participation across states and regions, and move the public from emergency compliance to internalizing awareness and behavioral change, is to engage a wide variety of community leaders in each municipality, county, and region to help amplify the messages and diversify the message delivery mechanisms, to reach the variety of audiences within an affected area. People must take the issue to heart and become personally invested, and the maximal involvement of leaders in the work of protecting and perpetuating healthy and resilient 'ōhi'a forests was a key goal of the strategy.

The third goal of the strategic plan was to ensure outreach coevolved with science. Our knowledge of this relatively newly identified pathogen, and the devastating effects on our keystone native tree was (and currently still is) increasing every week. However, there are still many questions and different avenues of research that must be pursued in the coming years. Outreach should increase the percentage of the public who understand ROD and are aware of the current actions and avenues of research. Outreach must also be clear about the evolving nature of scientific research, provide timely updates to keep the public informed, and communicate when new evidence requires a change in prevention methods.

The objectives of this approach to strategic planning were (and are) as follows:

1. **Expand capacity through creating and supporting new ROD outreach positions statewide**. Create and support six new full-time employees (FTEs): one FTE to coordinate statewide outreach and engagement on ROD, and five FTEs to work on ROD outreach and community engagement for their islands (one each for West Hawai'i Island and the islands of Maui, Oahu, Kaua'i, and Moloka'i/Lāna'i).

2. **Lend support to conducting public awareness assessments and surveys**. To better understand baseline awareness, support/ opposition, and messaging support by the general public statewide, focus groups should be conducted as a first priority. Professionally conducted focus group work helps to refine messages and actions for years to come. Assessment of awareness and progress in the coming years should be conducted via occasional benchmark surveys for quantitative measures, and engagement may be used for qualitative measures.

3. **Lend support for outreach and media**. To date, the production of outreach materials and media has been contributed in-kind and have not all been sufficient (albeit with several efforts, such as the Merrie Monarch Festival efforts, as clear examples of effective grassroots efforts). Further, new staff will need operational funding and support for production and delivery of outreach materials via radio, television, print, and other means. Purchasing of expensive television broadcast time is included in this need. Raising awareness about 'ōhi'a is a core need, as many residents statewide are not aware of the value of 'ōhi'a. Other items include statewide production and availability of decontamination kits for the public trail signage and other informational products that have been piloted on Hawai'i Island with great success.

4. **Lend support for partnerships and community engagement of different groups**. Support for staff and materials will enable engagement of more potential partners. There are many cultural leaders, educators, community leaders (e.g., in the hunting community), and others that could play a role in expanding outreach and engagement, yet it takes time and effort for staff to work with each group to build trust and ensure that ROD information and messages are understood and agreed upon.

5. **Lend support for curricula development**. One key avenue for reaching the public is via schools. There are several examples of existing curricula aimed at K-6th grade levels that help students better explore the value of 'ōhi'a. These curricula can be updated

and expanded to all grade levels including junior college, modified to incorporate ROD information and helpful actions, and aligned to the standards of science and traditional knowledge. Much of the existing curricula is not yet available online, so these and additional materials that could be developed will need to be made accessible to educators via the Internet. Appropriate curricula for intermediate and high schools could also be developed, aligned, and made available. Curriculum development and alignment with the newly produced educational standards are a specialized skill set; thus, this would be done via specialists in knowledge sharing. The outreach network and staff could work with school administrators and teachers to conduct teacher training and help them incorporate the curricula into their existing lesson plans and goals for student assessments.

6. **Lend support for organizing or participating in educational and cultural events**. Educational and cultural events help engage communities and provide a venue for increasing or perpetuating values. Each island has its own regular community and cultural events that can be used to share 'ōhi'a messaging, and thus far, this work is limited by staff time and availability of ready-to-use outreach displays and materials. Engaging the public in "Celebrating 'Ōhi'a" is being pilot tested on Hawai'i Island where awareness of 'ōhi'a appears to be high. However, based on the initial results of public surveys at Oahu and Maui hiking trails, resident awareness of 'ōhi'a and ROD is much lower on these islands. Awareness of 'ōhi'a is key to engaging the public in behaviors to protect 'ōhi'a.

7. **Maintain into perpetuity the monthly 90-minute "ROD Working Group."** ROD Working Group meetings had been held since August 2015 at the USDA Forest Service office in Hilo (i.e., The Institute of Pacific Islands Forestry) and via teleconference, accommodating (importantly) any and all participants. These meetings are typically attended by 15–35 individuals in person and 25–50 via telephone. As of 2018, approximately 350 individuals have requested to be on the ROD Working Group, and they receive periodic updates via e-mail. Participants include private citizens, managers, scientists, policy professionals, and communication specialists, among others, and the expanded list of participants includes nearly every affected organization and interest group in Hawai'i (and including several groups/individuals with expertise and interest from outside the State of Hawai'i). Notes for the ROD Working Group are routinely taken for informational purposes and disseminated among all on the e-mail list, along with any updates, information, and documents of the group. Nearly any group/organization that has an interest in the topic of ROD participates monthly, through the interactions of

the ROD Working Group and/or the CAT. The originally stated purpose/principles of the meeting are to "facilitate inclusive ongoing discussions/communication of all issues related to ROD and share knowledge on a regular basis among group members, their organizations, and the people of Hawai'i."

1.7 Transformational Change Realized

This initial example of the transdisciplinary approach taken to address the societal and ecological issues surrounding ROD in Hawai'i is emblematic of the way that, in theory, many environmental challenges and crises can be addressed more effectively around the world. In this one example, we briefly explored both the larger context and the flavor of "small things" (e.g., interpersonal relationships and the details of differing cultural norms) that occurred within the working team. These small, but important, interactions brought forth an important change, and in the following discussions and subsequent global stories of collaboration and transformation, you will see some very similar and powerful examples of these same phenomena, where a small impetus caused a large and impactful outcome for people and communities.

To those in Hawai'i, the ROD crisis is a multigenerational change, and indeed when you see the distance between today's situation and where the people of Hawai'i need to go to meet all of the current challenges of ROD, it does seem quite overwhelming. A stark departure from the seemingly overwhelming challenges of Hawai'i is offered up in Chapter 2, which takes us to Iraq. You will notice that in Iraq there are similar themes to Hawai'i, specifically in terms of a challenge that is paired with the drive and desire of people to achieve transformational change. The challenges in Iraq, however, make the ROD-related issues in Hawai'i seem relatively workable, by comparison. As you will see in Chapter 2, the tremendously resource-rich country of Iraq, after years of war and other conflicts, has been transformed from the "cradle of civilization" to a country with stunted development of most of the major sectors of their economy, with a multigenerational environmental crisis of monumental depth and breadth staring them in the face. Indeed, Iraq's heavy dependence on oil revenues in the past century has driven them even further into a situation where they do not have the diverse economy to withstand the destruction of two fundamentals, i.e., food security and access to safe and drinkable water, which have been exacerbated over the past few decades. Chapter 2 recounts these challenges in detail, with a few glimmers of hope, pointing toward a future Iraq continuing to struggle to overcome one of the world's worst conflict-related environmental disasters.

Disclaimer

The research, analysis, and other work documented in this chapter was fully or partially funded by the USDA Forest Service; however the findings, conclusions, and views expressed are those of the authors and do not necessarily represent the views of the USDA Forest Service.

2

Environmental Decision-Making within a Recovering War Zone: The Republic of Iraq

Mohammed A. Kalkhan

Colorado State University

CONTENTS

2.1 Background

Despite a history of conflict over the past three decades, Iraq has a tremendous peace-building capacity, and its people are now striving for redevelopment after years of war. Within the, now, strengthening civil society of Iraq, there are signs of stability and hope for the future for the governmental system that the people hope can respond more effectively to the needs of all Iraqis, based on a fundamental need for all societies: the principles of "the rule of law," respect for human rights, respect and support for academic endeavors and technological transfer, shared governance, as well as a desire and capability for collaboration with the international community. The focus of this chapter is to explain how these principles might now be successfully applied to restoring environmental conditions in a devastated country like Iraq, and how the fundamental principles stated above can be strengthened "through the work," as key environmental restoration projects in Iraq proceed, leveraging the power of the international community to heal a traditionally peaceful way of life in Iraq.

Iraq is a resource-rich country, but years of war and conflict have hindered sustainable development of agriculture, water resources, mining, energy, and tourism, as well as health and education systems. Increases in oil production, a primary source of national wealth, have not kept up with the decline of oil prices worldwide over time, thus resulting in large budget deficits and an economic recession, leading to Iraq's diminishing success in relying on oil revenues. Consequently, food security and access to safe and plentiful water are persistently problematic issues for a large part of the population in Iraq. Consequently, the government has been working to encourage foreign investment into environmental restoration of watershed functions, renewable energy, and other natural resources sectors (e.g., mining).

Iraq faces numerous environmental problems including severe pollution, intertwined with a host of other issues (e.g., dust–sand storms, water quality and quantity, oil products and fires, and industrial discharges into the water and air), principally because of decades of war, international sanctions, and general mismanagement by the Saddam Hussein regime. As outlined by the United Nations, nearly a decade and a half ago, "A major threat to the Iraqi people is the accumulation of physical damage to the country's environmental infrastructure" (UNEP 2005).

Current environmental priorities for postwar Iraq are restoring water and sanitation systems, cleaning up possible pollution hot spots and waste sites, and providing information to residents on minimizing the risk of exposure to depleted uranium, which was used by coalition forces in its ordnance. The specific tasks needed to clean up hazardous wastes and emissions in Iraq involve the improvement of water management and the restoration of the country's premier ecosystem, the Mesopotamian Marshlands, of which 90% were drained on Saddam Hussein's orders after the 1991 Gulf War. Lack of investment in water and sanitation systems has also led to increased pollution and health risks, while electricity cuts have often shut down pumps that remove sewage and circulate fresh water throughout the marsh areas (UNEP 2005).

The destruction of military hardware and factories during Iraq's various wars has also released heavy metals and other dangerous substances into the air, soil, and water. Smoke from oil-well fires and burning oil trenches (a battlefield tactic used by retreating forces during conflicts) caused localized air pollution and soil contamination. Heavy bombing and the movement of large numbers of vehicles and troops have also degraded the ecosystems of the area. Consequently, when the desert's hard-packed surface is disturbed, the underlying sand is exposed and often erodes or blows away (Figure 2.1). Relatedly, the United Nations Environmental Program (UNEP) has recommended an assessment of any environmental contamination that may contain traces of depleted uranium weapons that remain after the 1991 Gulf War (such as in blowing dust/sand), based upon their previous studies of the impact of similar conflicts on the former Yugoslavia, Albania, and

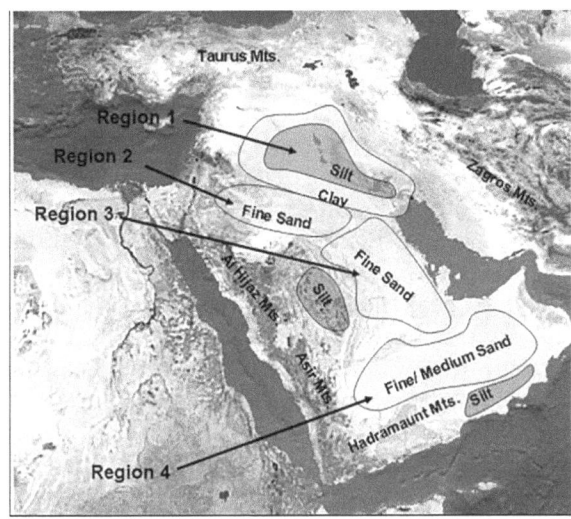

FIGURE 2.1
(See color insert) Southwest Asia dust source regions, and the areas of impact within Iraq. (In Anderson 2004 and adapted from COMET 2003.)

Palestinian territories; to date, they have been unable to do their own comprehensive study of Iraq on the ground.

2.2 People, Geography, and Climate

The human landscape of Iraq is one of multiethnic and multireligious peoples, with Islam, Christianity, Yazdanism, Zoroastrianism, Shabakism, Judaism, Mandaeism, Bahā'i, Ahl-e Haqq-Yarsanis, Ishikism, and numerous other religions all having a presence in the country. Shia Islam is the main religion in Iraq, followed by 60%–65% of the population, whereas Sunni Islam is followed by 32%–37% of the people. Many cities throughout Iraq have been areas of historical prominence for both Shia and Sunni Muslims, including Najaf, Karbala, Baghdad, and Samarra. These geographies and the belief systems of the people living there factor into the traditions of the particular area, and thus the current focus topics for environmental restoration.

Iraq predominantly lies between latitudes 29°N and 38°N, and longitudes 39°E and 49°E, in the Middle Eastern Region of the world. Spanning 437,072 km² (168,754 mi²), Iraq is the 58th largest country in the world. Iraq is comparable in size to the state of California and is somewhat larger than the South American country of Paraguay. Iraq borders the Persian Gulf and six other countries, which include Kuwait, Saudi Arabia, Syria, Jordan, Iran, and

Turkey (Figure 2.1). Most of Iraq's land is desert, with most of its mountains found in the northern region of the country, along the borders with Iran and Turkey. Iraq mainly consists of desert; however, the major rivers (Euphrates River and Tigris River) support what amounts to a very vast area of fertile alluvial plains in the country, with these rivers carrying about $60,000,000\,m^3$ $(78,477,037\,yd^3)$ of silt annually to the delta.

The mean elevation of the country is 1,024 feet, with the highest point at Cheekha Dar (translated as "Black Tent"), located in Irbil, Iraq, at 11,847 ft (3,611 m), whereas the lowest point is in the Persian or the Arabian Gulf, at sea level. Iraq's highest peaks are in the northern part of the country, in and around the Erbil Governorate. These peaks are part of the Alpine system, which runs through Turkey, Iraq, Iran, and Afghanistan and eventually joins the Himalayas. Iraq has a small coastline measuring only 58 km (36 mi), along the Persian Gulf. North of this coastal area and along the Shatt al-Arab, there are many marshlands that were among those drained in the 1990s war period, which we will delve into in detail with regard to wetland restoration, later in this chapter.

Most of Iraq has a hot and arid climate with subtropical conditions (Figure 2.2). Summer temperatures average above 40°C (104°F) for most of the country, frequently exceeding 48°C (118.4°F). Winter temperatures infrequently exceed 21°C (69.8°F), with daily maxima at roughly 15°C–19°C (59.0°F–66.2°F) and nighttime lows of 2°C–5°C (35.6°F–41.0°F). Typically, precipitation is low; most places in Iraq receive less than 250 mm (9.8 in) of precipitation annually, with maximum rainfall occurring during the winter

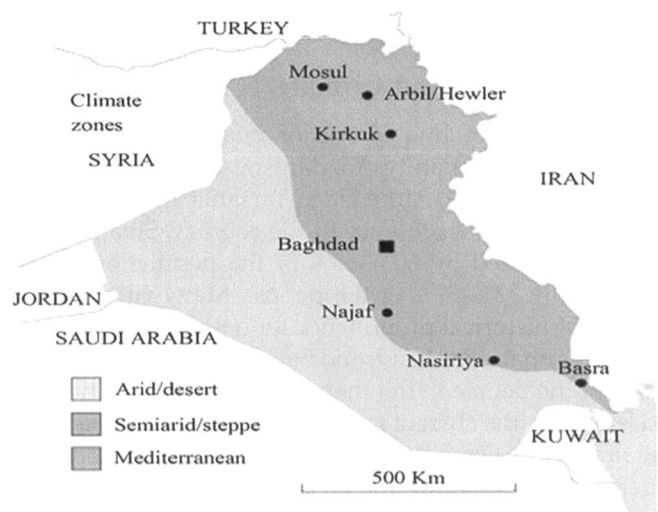

FIGURE 2.2
(**See color insert**) Generalized environmental conditions in Iraq and climate zones. (UNEP 2005.)

months. Rainfall during the summer is extremely rare, except in the far north of the country. The northern mountainous regions have cold winters with occasional heavy snows, sometimes causing extensive flooding.

2.3 Geopolitical, Scientific, and Technology Challenges

Over the past 25 years, Iraq has incrementally become much more isolated from the world community of nations because of UN sanctions, the Iraq–Iran war, and the Gulf war. These events prevented Iraq from keeping up with the rapid advances in science and technology occurring in the rest of the world. Specifically, Iraqis have incrementally become lacking in the resources necessary for capacity building and development, and institutions within Iraq do not always have the personnel to conduct technological work nor provide internal technical transfer of scientific information, which is necessary to understand the extent and scope of environmental conditions and the societal impacts of those conditions, nor is there sufficient internal capacity for the needed application of concepts related to environmental sustainability, scientific inquiry, or technological capabilities for restoring landscapes, agriculture systems, and water resources. A major example of this deficit of expertise and technological transfer is in the use of geospatial information and the technology necessary for environmental and ecological forecasting (i.e., remote sensing, geographic information systems, the use of drones, and other technologies).

2.4 Socioecological and Ecosystem Threats

Among several socioecological challenges that the people of Iraq are continuing to face is the serious shortage of a sufficient and safe water supply, particularly in terms of agricultural demands. This shortage of water specifically hampers efforts to expand and improve agricultural and land management capabilities in the country in a healthful and sustainable manner. Iraq is a land driven by its two major rivers, the Tigris and Euphrates, and historically a very fertile land, the renowned Mesopotamia, with the convergence of these two rivers once producing such rich and fertile soils and supplies of water for irrigation across that this vast area of the landscape was capable of supporting a burgeoning civilization that emerged around these rivers, among the earliest known non-nomadic agrarian societies of the world. It is because of these fertile conditions of abundance that the Fertile Crescent region, and Mesopotamia in particular, is often referred to as the cradle of civilization. The period known as the Ubaid period (ca. 6,500–3,800 BC) is the earliest known period on the alluvial plain of Iraq, and likely even earlier periods existed there, where a complex and productive civilization prospered, which is obscured by areas far under the alluvium that are not available to archeologists any longer. It was within this highly productive area during the Ubaid period that the early path toward urbanization began. Agriculture and animal husbandry were widely practiced in sedentary communities prior to this period, particularly in Northern Mesopotamia, as intensive irrigated hydraulic agriculture began to be practiced in the south. By contrast, Iraq now suffers from massive effects of desertification and soil salination, due in large part to thousands of years of agricultural activity, such that water and plant life are now sparse. Water projects have ensued over the past several decades in Iraq, and prior to the Iraq invasion by the United States in 2003, Saddam Hussein's government instituted massive water control projects, which drained the inhabited marsh areas east of An Nasiriyah and other locations (Figure 2.3), drying up or diverting

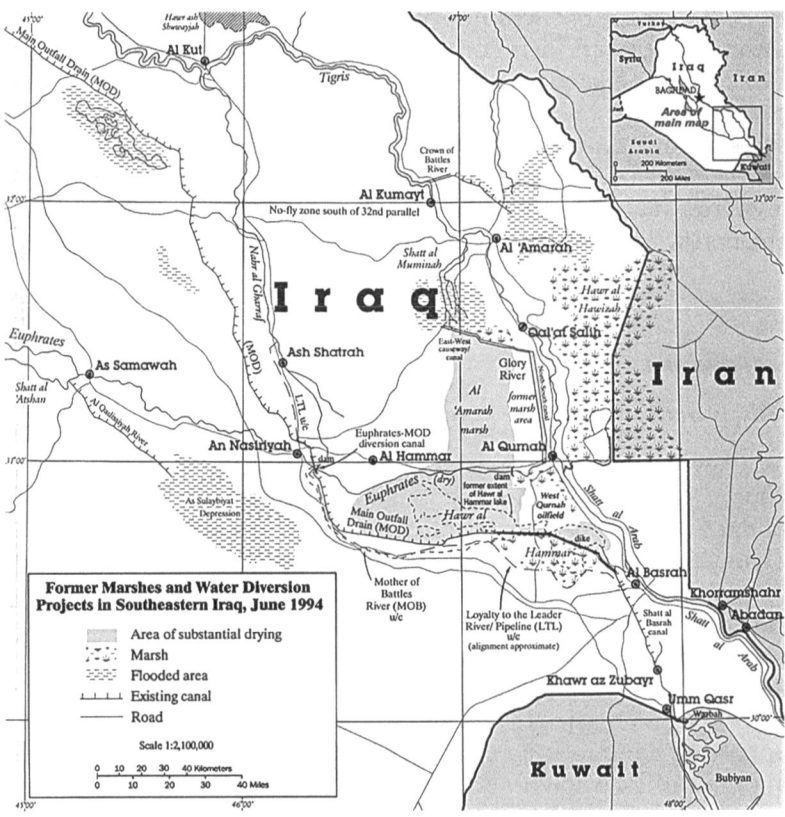

FIGURE 2.3
(**See color insert**) Iraq's marshlands. (Cited from Google Maps incorporated from UNEP.)

streams and rivers. Consequently, historically resident Shi'a Muslims in the marshlands were displaced under the (Ba'athist) regime of Hussein. The consequential destruction of natural habitat for specific wildlife species, and the way of life for people of the marshlands, posed serious threats to the area's people (e.g., Kubba 2011; Al-Mudaffar et al. 2016) and ecosystem functions, impacting everything from wildlife populations, to plant life, to hydrologic functions, and severely altering supplies of potable water for communities.

2.5 Restoring the Gift of Water through Marshlands

As alluded to previously, Iraq is extremely dependent upon water for its survival. Yet, Iraq is a "riparian country," within the Tigris–Euphrates River basin, and hence, the development of the country has always "followed the water," as is the case in other similar countries in the region, and indeed around the world. The water demands within the country have continuously increased since the original people of the area existed, and due to population growth, economic development, and environmental priorities the country has become even more dependent upon the precious water resources of the two rivers (Al-Ansari 2013). Consequently, the entire region is currently facing extreme water shortages, and the concomitant problems that accompany water scarcity burden the region. Continued pressures from population growth are expected to increase the demand for the foreseeable future, with climate change as an exacerbating force. Yet, until 1970, Iraq was considered relatively rich in its water resources, from a regional standpoint. However, in the mid-1970s, Syria and Turkey started to build dams on the Euphrates River, and subsequently the river volume flowing south toward the Persian Gulf, through Iraq, decreased, the result of the numerous associated impoundments. A number of water projects were also proposed just prior to the period of wars, beginning in the 1990s, and the Iraqi Government is now attempting to reinvigorate them and speed up building those projects as rapidly as possible to meet demand. Indeed, from 1970 to 1990, the most rapid period of hydrologic project development in Iraq's history occurred. Today, the demand for water is ever-increasing in Iraq, with an annual demand for water of $1.002\,km^3$, partitioned as 0.5271 and $0.475\,km^3$ from the Tigris and Euphrates basins, respectively. Despite shortages now, the average water demand in 2020 will increase even more, to $42.844\,km^3/$ year for the Tigris basin and $29.225\,km^3/year$ (total $72.069\,km^3/year$) for the Euphrates basin, whereas water availability will decrease to $63.46\,km^3/year$. This means that the overall water deficit will be $8.61\,km^3/year$ (FAO 2015). In addition to these projected deficits, the combination of climate change, population growth, and limited environmental awareness seriously limits the options for the Iraqi national water policy, further impeding solutions to the troubling associated societal issues, such as low crop production and shortages of basic water supplies for drinking water.

Prior to the present lack of water resources, in 1993, the maximum recorded area of agriculture was 9.87 million ha. In addition, the area of mean arable land and permanent crops was approximately 5.05 million ha, representing 11.5% of the total land area of Iraq, which for Iraq is high. Compared to many other countries around the world, the proportion of arable land in Iraq is relatively small, as a result of the scarcity of land suitable for cropping in Iraq. Although more than half (53%) of the arable land in Iraq is supported by precipitation, with nearly all of it in the northern uplands of Iraq, most of the agricultural production for the country comes from the more intensively cultivated areas of the irrigated plains. This results in the agricultural production of Iraq being extremely dependent upon the health and functions of riverine and riparian ecosystems in the lower reaches of the watershed, in the alluvial floodplain and wetland areas of southern Iraq. Because of soil salinity, high water tables, poor farming practices, and the precarious political circumstances over the past few decades, it is estimated that only 3–5 million ha of land were actually put into agricultural production annually, for the entirety of Iraq.

The marshlands are closely linked, both physically and societally, with the rivers of Iraq, and their socioecological connectivity is a crucial element of the future survival for the people and the ecosystems of Iraq. The marshlands are a big part of the story of hope for environmental restoration in Iraq, and these transitional ecosystems (with both upland and aquatic characteristics) were once an extensive natural wetlands ecosystem, developed and managed by humans over thousands of years in the Tigris–Euphrates basin. The marshlands of Iraq once covered 15,000–20,000 km² (5,792–7,722 mi²), about the size of the country of Albania. According to the UNEP and the AMAR Charitable Foundation (www.amarfoundation.org; checked September 15, 2018), between 84% and 90% of the marshes have been destroyed since the 1970s. In 1994, 60% of the wetlands were destroyed by the Hussein regime's policies, primarily to allow for military access and greater political control of the native Marsh Arabs' lands. Canals, dikes, and dams were built such that the water from the Tigris and Euphrates rivers was routed around the marsh areas, instead of allowing water to move slowly through the marshland as it would naturally. After part of the Euphrates was dried up due to this rerouting of water, a dam was then built so that water could not back up from the Tigris, further isolating the former marshland hydrologically. Some marshlands were also burned and a network of pipes were buried underground within them to shunt away water more efficiently from the marshes, causing them to dry quickly and thoroughly.

The drying of the Iraqi marshes led to the disappearance of most of the salt-tolerant vegetation of the region, which is native to the area; plankton-rich waters that fertilized the surrounding soils were lost; 52 native fish species were extirpated; and large animals such as the wild boar, red fox, buffalo, and water birds of the marshes disappeared. The wetlands in the deserts of southern Iraq provided habitat to many hundreds of thousands of birds and represented a unique archeological monument that dates back 5,000 years (UNEP 2011). Thirty-three Iraqi wetlands were included on a 1993 provisional list of wetlands of international importance in the Middle East, noted for their support of a substantial number of at least seven species of mammals and birds that are listed in the IUCN Red List of Threatened Animals. The Iraq marshlands are specifically of international importance as a staging and wintering area for more than 60 species of waterfowl and 9 species of birds of prey. The tremendous cultural significance of the Iraq Marshlands is also quite clear, and the wetlands have created a home for the Ma'dan, or Marsh Arabs, for at least the past 5,000 years. Thus, it is no wonder that the marshes are considered by some scholars as the Biblical Garden of Eden, which have suffered greatly over the last half century, drained repeatedly since the 1950s (Al-Mudaffar et al. 2016) and further damaged by drought, desertification (Chenoweth et al. 2011), and physical destruction from violent conflicts. Consequently, by 2003, the marshes had shrunk to a mere 10% of their original size, with much of their socioecological functions eliminated. Assistance from the UNEP, however, has provided some new hope for the ecosystem, gradually returning the marshes to their former structure and ecological functions. Since 2004, the United Nations Development Programme (UNDP) has also been working with Iraqi communities and the Government of Iraq to restore the social and ecological elements of the marshlands to their former glory. These collaborative partners have trained Iraqi decision-makers, monitored marsh conditions, and worked in partnership with local communities to develop a plan for professional management of the marshes' sustainably.

After sustained collaboration and progress, a major breakthrough came in July 2016, when the marshes were granted World Heritage status by the United Nations Educational, Scientific, and Cultural Organization. This momentous event was a result of key pursuits by a number of the partners involved, one of which is within the UNEP and that played an important role in the negotiations for the new status of the marshes. This new designation brings with it new funding and international attention that is much needed in the area, and the new status further encourages the Marsh Arabs and others of the region to preserve their environmental and cultural heritage. Local community leaders greeted the announcement with enthusiasm, with the general belief that "having our marshes on the World Heritage list is a big dream come true for us," said "Sabon," a local fisherman and native of the Marshes. "Our life will be better, income will be better, and tourism will prosper." Sabon was born in the central region of the Marshes in 1965, but he left with his family in the early 1990s, after the regime of Saddam Hussein had begun draining the wetlands. In the early 1990s, Sabon and his family had to move to the outskirts of Baghdad, where Sabon found work as a farmhand. But in 2005, when the marshes started to come back to life, Sabon and his family returned to the area. Now, in addition to fishing, people also make money by operating boats for tourists.

"There are a number of things we need to do," said Mr. Jassim Al-Asadi, director of Nature Iraq, a nongovernmental organization (NGO) and UNEP partner; "we need to carry out further studies on the marshland ecosystems, assess how human activity is affecting the local environment, and get local businesses involved in schemes to maintain and protect the environment." Furthermore, Mr. Al-Asadi stated, "Our support to the Iraqi Marshes is just one example of our work to promote integrated ecosystem management—an approach that aims to help ecosystems meet both ecological and human needs. Critically, this kind of work can help to prevent new [emigration] flows of environmental migrants, like fishermen of the marshland" (UNEP 2016).

As is evidenced by many other historical circumstances, environmental factors have long been a driving factor in global migration, as people flee harsh and deteriorating conditions in search of better lives and livelihoods. According to the International Organization for Migration, the world could see up to 1 billion environmental migrants by 2050 as a result of similar circumstances as occurred in Iraq; however, in Iraq it appears as though there may be an improved set of circumstances, i.e., the restoration of the marshlands, which may alleviate one source of this ecologically driven societal and national security problem.

2.6 Collaborative Lessons from Marsh Restoration Progress

Restoration of the Marshes of Iraq is a good example of the complexity of environmental restoration challenges in Iraq. These efforts and the broader assessment of the surrounding landscapes and those that are contributing water flow to the Marsh areas, and beyond, have been undertaken during the past several years, with the UNEP and other international organizations providing capacity building assistance and training for the Iraqi Ministry of Environment and Health to conduct assessments of many polluted sites. Much has been learned from the successes within the marshlands of Iraq, which may lead to additional progress on other environmental challenges facing Iraq. In addition to the UNEP Teams, other complementary contributions from Iraqi counterparts have occurred, including field visits and laboratory analyses. These activities are the bases for the cleanup and recommendations for remediation of the identified contaminated sites throughout Iraq, involving training courses that underline the importance of establishing a link between the Iraqi experts within the country with the international experts that are in Iraq to assist, thus increasing the likelihood of effective assessment outcomes. Some of the confirmed contaminated sites, where further analysis is needed, have been specifically identified at this point as follows:

1. Al-Mishraq Sulphur State Company
2. Midland "Al-Doura" Refinery Company
3. Al Suwaira Contaminated Seed Store
4. Country-wide oil fires and spills area
5. Military vehicle stockpiles

The UNEP teams observed that Iraqi industries are primarily concentrated along the major rivers of the country, with environmental problems that are very similar to those experienced in Central and Eastern European countries, making the rivers a major focus of remediation activities in Iraq.

For the UNEP Iraqi Marshland Projects, the team recognize interlinking issues such as cultural, political, social, and agricultural factors, and so they integrated this experience into their thinking about how the unique lifestyle of the Marsh Arabs (Figure 2.3) fits within the larger restoration plans for Iraq. Since the initiation of the re-flooding of the Marshlands, a number of positive ecological and sociological effects have been observed, as described in detail previously. These areas, however, are still facing enormous risks brought about by the surrounding landmines, especially those that remain unsurveyed. Moreover, the Marshlands are facing other types of hazards, such as unexploded ordnance, (i.e., bombs, missiles, rockets, and other unknown dangers associated with battlefield operations). The presence of such materials is one of the factors preventing the full restoration of the Marshlands; at this point in time approximately 10% of the Marshlands have been returned to a functional ecosystem, i.e., the Al-Hawizeh/Al-Azim Marshes. In addition to the above-described refugia for biodiversity being provided by the Marsh areas, they also act as excellent reference areas for future marsh restoration activities, making them eligible for international conservation status. The UNEP-Iraqi collaborative Team took the lessons from the Iraqi Marshlands Project to heart and transferred that knowledge to the broader assessment work for the entire country, specifically in terms of collaboration and partnership.

The UNEP projects are not only "ecosystem restoration projects," they are important infrastructure projects where previous hydrologic disturbances have specifically impacted other landscape components that are important to society, such as resulted from dam construction, as follows:

1. Major alteration to the human water supply
2. Reduced capacity for the people of the Al-Hawizeh/Al-Azim marsh areas to carry on traditional daily activities
3. Negative impacts on water quality for drinking purposes

The collaborative team noted some additional issues affecting the Marshes, which affected everyday life of Iraqis, including the border dikes between neighboring countries (i.e., Iran) that impede normal hydrologic processes from occurring. The indispensable (i.e., positive) influence of the Centre for the Restoration of Iraq Marshlands (CRIM) within the Ministry of Water Resources of Iraq is also focusing on several other key "big picture" issues as follows:

1. A need for developing the local scientific knowledge base to help guide operational action planning
2. The importance of building national, institutional, and technical capacity
3. The critical role of catalyzing regional dialog and cooperation to allow for success. Importantly, the team observed how the context of the marsh restoration and protection work provides knowledge and practice in the use of tools for fostering peace and building cooperation among international partners

In summary, the Iraqi Government water control projects drained most of the inhabited marsh areas east of An Nasiriyah, mainly by drying up or diverting the feeder streams and rivers. A once sizable population of Marsh Arabs, who inhabited these areas for thousands of years, was displaced due to political factors of uncertainty and as retribution for their uprising against the Hussein government in the 1970s and 1980s. Accordingly, the historical destruction of natural wetland habitat types now poses serious threats to the area's cultural heritage, wildlife populations, supplies of potable water, air and water quality, and soil stability (i.e., erodibility) and quality (e.g., salination), and increases the pace of desertification in the region. These factors are, from an ecological point of view, quite important, and at the same time there is tremendous sociopolitical pressure to develop the Tigris and Euphrates river systems, yet this is very contingent upon agreements with upstream areas, such as Turkey, as well as other societal factors within Iraq.

2.7 Broader Socio-Environmental Impacts of War

There has been much debate about the war in Iraq, but not about the considerable environmental damage that the war caused. During the 1991 war in Iraq, devastating damage was done to the oil industry in Kuwait. Iraqi forces destroyed more than 700 oil wells in Kuwait, spilling 60 million barrels of oil across the landscape. Over 10 million m^3 of soil was still present and contaminating the land, as late as 1998. A major groundwater aquifer,

40% of Kuwait's entire freshwater reserve, remains contaminated to this day. Ten million barrels of oil were also released into the Persian Gulf, affecting 1,500 km of coastlines and costing more than $700 million to clean up. During the 9 months that the wells burned, average air temperatures fell by 10°C (50°F) because of reduced light from the sun. Total costs of environmental damage from the 1991 war are estimated at $40 billion. Since Iraq has the second largest proven oil reserves of any nation on earth, the potential environmental damage caused by the destruction of oil facilities, during any war in the country, is enormous. Other environmental effects of the 1991 Gulf War included the destruction of sewage treatment plants in Kuwait, resulting in the discharge of over 50,000 m^3 of raw sewage every day into Kuwait Bay.

Iran also plays a major role in the current circumstances in Iraq, with a population of 50 million (compared to Iraq's 17 million people) and at war with Iraq in 1980 during the Iranian Revolution. By the summer of 1982, Iraq was on the defensive and remained so until the end of the Iran–Iraq War in August 1988. The death toll for the Iran–Iraq War, overall, was an estimated 1 million people for Iran and 250,000–500,000 people for Iraq. However, in the west, it often seemed, even at the time, like a forgotten war. Far more attention from the west was paid to later conflicts, such as Iraq's invasion of Kuwait in 1990 and the U.S.-led invasion and occupation of 2003, which overthrew Saddam Hussein's Regime and dramatically changed the political map of the Region. The effects of all of these wars were devastating for the people of the Region, especially Iraq, including social impacts, public health (mortality, cancer, respiratory problems, stress due to lost family members), as well as economic and environmental impacts.

Specific weapons used in Iraq also created unique environmental damage, such as those of extreme concern like depleted uranium projectiles. Depleted uranium is very dense and is used in projectiles designed to pierce armor, reinforced bunkers, and other similar targets. Depleted uranium projectiles create fragments and dust, which release uranium oxide into the air upon impact. Estimates of the amount of depleted uranium used by allied forces in the first Gulf War range from 290 to 800 tons. Decontamination of this type of battlefield contamination requires removal of contaminated soil and treatment of the soil as radioactive waste. Thousands of hectares of Iraqi land could be contaminated in this manner and just 200 ha decontaminated at a U.S. Army proving ground was costly, at approximately 5 billion dollars (United States). The UNEP (2007) stated that "According to a 'threat paper' on Kuwait produced in secret by the UK Atomic Energy Authority and subsequently leaked, 50 tons of deteriorated uranium (DU) inhaled could cause up to half a million additional cancer deaths over several decades, a calculation based on International Committee on Radiological Protection risk factors." Internal DU exposure is acknowledged to cause kidney damage, cancers of the lung and bone, respiratory disease, neurocognitive disorders, chromosomal damage, and birth defects (Faa et al. 2018).

Battlefield conflicts also pose serious threats to the biodiversity and landscape ecological processes of the region. Data on Iraq's biodiversity is limited, with only a little information on fish, amphibians, and reptiles. No major surveys have been conducted since 1979, but Iraq's wetlands have been of major international significance, especially for wildfowl, and studied much more extensively than other areas, as outlined in the previous sections.

2.8 Iraq's Continuing Struggle with "Conflict Pollution"

As illustrated previously, the effects of the war in Iraq were devastating to the landscape, but most especially for the people of the Region, including but not limited to societal and familial disruptions, physical violence, mortality of civilians and combatants, cancer, respiratory problems, economic distress, and environmental pollution. Since the U.S. invasion in 2003, the violence in Iraq continued, spawning sectarian and insurgent attacks that were responsible for the majority of residual and current violence, as it is just today (slowly) coming under control. Most attacks targeted civilians based on their ethnicity or religion, or perceived affiliation with the new Iraqi government or occupying forces. At many points along this path many believed that Iraq was on the brink of civil war. Currently, determining the total number of deaths resulting directly and indirectly from the armed conflicts in Iraq is challenging, and the war has caused the destruction of so much of the health infrastructure and health information systems in Iraq, systems already weakened by UN sanctions in the decade prior that organizing a system of basic health care for the public is extremely difficult. A UN study of deaths due to violence since March 2003, determined that approximately 759 deaths occur per day (range 456–899 people per day) as a result of postwar deficits in environmental conditions and health-care support of the public (UNEP 2003).

In 2015, while Iraq was still recovering from the environmental impacts of both Gulf wars, it faced a new socio-environmental threat caused by internal conflicts with the Islamic State (a self-styled extremist movement, also known as ISIS, ISIL, or, by its Arabic name, Daesh). While Iraq was still recovering from the environmental impact of both Gulf wars, described above, it was also facing new environmental problems caused by this newly activated conflict; since the uprising of the Islamic State in June 2014, fierce battles took place in and around cities and industrial areas of Iraq, affecting the already precarious environmental situation, again resulting in oil fires, such as in Figure 2.4.

Despite Iraq being ravaged in recent years by cycles of warfare, a growing refugee crisis, crippling sectarianism, and the violent spread of ISIS, Iraqis have made some progress in building their government, specifically by

FIGURE 2.4
An oil fire during the first Gulf war, in 1991 (www.publichealth.va.gov/exposures/gulfwar/sources/oil-well-fires.asp; checked September 15, 2018).

approving a constitution to replace that of the Saddam Hussein era, and holding successive elections for parliament and provincial governments. Despite this progress, current governing institutions remain weak, and corruption and poverty are prevalent. The ISIS threat and rising violence compelled U.S. military advisors to return to the country in 2014, after having withdrawn in 2011. The continued weakness of governance in Iraq, along with ISIS' seizure of much of northwestern Iraq and adjacent parts of Syria posed a long-term challenge to sociopolitical stability at that time, which continues to hamper the recovery of Iraq. Heavy fighting in and around the Baiji oil refinery and attacks on other industrial installations led to recent releases of a range of hazardous substances into the environment, affecting the soil and groundwater. During the ISIS resurgence, battles took place near Kirkuk's oil fields, as ISIS focused their attacks on oil and gas installations, thereby increasing the likelihood of chemical incidents and subsequent environmental contamination and civilian exposure to hazardous substances.

Having learned lessons from the legacy of previous conflicts in the country, Iraq's Ministry of the Environment (MoE) diligently developed an assessment strategy to monitor the impact of conflicts, to speed up remediation work so as to limit air and soil contamination, as well as water pollution (Figures 2.5–2.8). According to the Minister of Environment, Mr. Qutaiba al-Jubury, "the Ministry will work to fight pollution by adopting techniques for monitoring, control, sensing and the application of effective preventive processes and procedures, … [aiming to] combat pollutants that threaten the environment and public health, in addition to those related to landmines and explosives planted by terrorists in residential neighborhoods, homes, and roads to impede the progress of security forces." Mr. Al-Jubury has accused ISIS of intentionally polluting water sources with oil waste and toxic chemicals, and destroying agricultural land, leading to desertification, economic losses, and threatening the food security of Iraq. It appears that ISIS has adopted a strategy of using environmental damage as a weapon of war through the

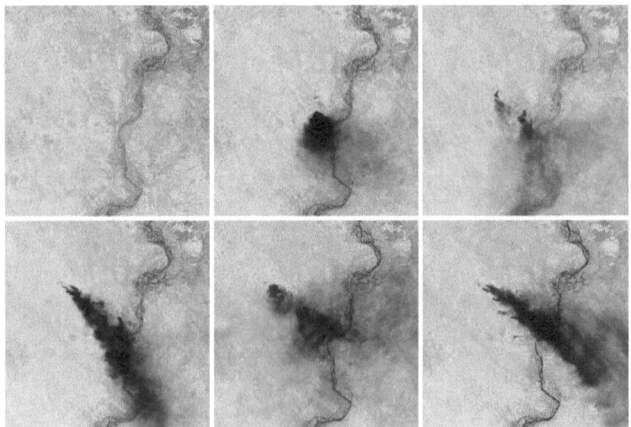

FIGURE 2.5
NASA satellite imagery of Qayyarah oil field burning (Saladin Province, Iraq) in the summer of 2016 (NASA Landsat 8). (From Schlanger 2018.)

FIGURE 2.6
Ground-level view of oil fires in the northern Saladin Province, in the Hamrin Mountains, Iraq, on September 30, 2017. (Photo by Mr. Hossein Velayati; from Schlanger 2018.)

"deliberate contamination of rivers, lakes, and streams with toxic waste and oil contaminants."

To address the problem of "conflict pollution," the MoE was reported to have developed a formal environmental assessment of such affected areas. Mr. Luay al-Mukhtar, Director of the MoE's Department for the Regulation of Chemicals and Polluted Sites, noted that this approach of monitoring and rapid response is an effort to "…prevent the spread of disease and to safely dispose of waste. As part of this project, Ministry teams will perform laboratory checks and analysis of water, soil, and air in every area that has been cleared and secured from ISIS control."

FIGURE 2.7
An aerial image of the Qayyarah oil field fires, December 7, 2016. (From Schlanger 2018.)

FIGURE 2.8
"Mahmoud Ali" (name changed to protect his identity) stands in the entryway to his house in Qayyarah. Oil fires burned outside his home for months. (From Schlanger 2018.)

Iraq's monitoring program outlines the importance of recording and assessing environmental damage during a conflict. Its focus is on assessing pollution risks to "…provide quick treatment and the right measures to contain the effect of conflicts or disasters on the environment in general" and to "curb the threat to human health posed by epidemics, diseases and deadly pollutants. Such activity and technical operations require continuous efforts on the ground for a sufficient period to draw conclusions and compile valid reports" (Zwijnenburg and Wim 2015). Naturally, effective assessment and monitoring requires technical capacity, expertise, and funding, as well as the security necessary for work to be undertaken safely and completely. These factors are not always present in conflict settings, which can delay the identification, assessment, and cleanup of a range of hazardous substances.

In cooperation with Iraq's Ministry of Science and Technology, the MoE has established programs to undertake environmental assessments on former military facilities from the Saddam Hussein era. Planned in cooperation with international organizations, the Ministries intend to "...check, monitor, and destroy any residue or waste found at contaminated sites" and, according to a MoE spokesperson, they are ready for additional efforts in "analyses of the environmental impact of violence." The aim of the program is to "...protect residents from any pollutants that could put their health and the health of their children at risk."

In addition to ongoing public health concerns over damage to Iraq's national physical infrastructure and vital ecological goods and services, such as water supply and sewer-watershed integrity, the World Health Organization (WHO 2017) has also highlighted technological and chemical hazards related to the conflict. They outlined that the conflict would increase the risks of releases from industry, particularly the oil industry, with consequences for public health. They were especially concerned about the security of former chemical weapons complexes such as at Muthanna, near Baghdad. In June 2014, ISIS militants took control of the Muthanna chemical weapons complex, approximately 70 km from Baghdad. The complex is believed to have held 2,500 rockets filled with sarin, about 180 tons of sodium cyanide, as well as munitions and storage vessels containing residues of mustard agent, all sealed in two bunkers. The sarin rockets were old, degraded, and thought to be unusable. The status of the sodium cyanide was unknown. It is believed that the chemicals would be extremely hazardous to anyone attempting to handle them.

This and other specific toxin situations clearly illustrate how conflicts increase the risks of a variety of different hazardous chemicals being released into the environment, with consequential impacts on human health. Potential hazardous event scenarios include the prior examples with regard to oil or munitions, as well as chemical releases from damaged factories, warehouses, and workshops, or the deliberate release of highly toxic chemicals. One such situation arose in 2007, in which trucks carrying chlorine cylinders were bombed, releasing chlorine gas into the immediate environment (Cave and Fadam 2007).

In summary, two very toxic sources of pollution exist in Iraq because of industry and the legacy of battlefield conflicts: oil and weapons. Because the main industry in Iraq is oil, damage to an industrial site or infrastructure typically hits oil drilling sites, refineries, and pipelines. Whether deliberate, sabotage, or accidental, the risk of environmental contamination and the possibility that drinking water and food source pollution is extremely high in Iraq. Crude and/or refined petroleum compounds are all very noxious, such that when there are installation fires they produce plumes of pollutants with irritant smoke, representing another potential hazard to human health and the environment.

2.9 Interventions to Address Recent Conflict Pollution

Given the complexity and the scope of the pollution problems in Iraq, it is evident that the People of Iraq are in need of support from the global community, and a number of partners have come forward to help, namely, the United States, the European Union, Canada, and Japan. A fundamental need that also lends itself to exacerbating the circumstances in Iraq is tied to the lack of adequate opportunities for jobs, education, and public health services, as well as a lack of effective policies for water, agriculture, natural resources management, and environmental protection. Consequently, international capacity building and development efforts are a principal focus in Iraq today, which includes technology transfer.

With limited information coming out of areas of Iraq that are, now, just recently, no longer under ISIS control, the risks to civilians from pollution generated by these most recent conflicts (in addition to the pollution from older conflicts) in these areas are currently unclear. Security concerns and a range of other humanitarian and financial priorities are limiting environmental research and attention at this time. Research by the Iraqi MoE could be a useful first attempt of hazard identification, exposure assessment, and risk characterization, leading to the identification of priority intervention areas, ideally with international support by relevant expert organizations.

As priority interventions, public access to sites of a chemical release should be controlled to prevent (further) exposure, and, if possible, any chemical releases should be contained. It is important that any chemically contaminated patients are decontaminated before they enter a health-care facility to prevent staff and other patients from being affected by the chemicals from the exposure incident. Health authorities should provide information to health-care staff and the public about the hazards of high-risk chemicals and the appropriate action to take. Health-care staff should also be trained in the recognition and management of symptoms of chemical exposures (WHO 2017).

2.10 A Legacy of Health Hazards

Three decades of conflict have led to widespread damage to Iraq's environment, its people, and its government's capacity to monitor and address problems.

Following the 2003 invasion, the UNEP published desk studies and post-conflict environmental assessments on the varieties of impacts of war on Iraq.

In addition to the environmental impacts, the reports identified long-term health risks caused by attacks on industrial infrastructure and their subsequent looting, contaminated military scrap metal, and the collapse of governmental oversight of sensitive materials and waste management. The UNEP urged Iraq and the international community to deal with these hazards swiftly, but efforts were undermined by the lack of a functioning government, insufficient capacity, and a general lack of expertise and the resources to identify, assess, clean up, and monitor problematic sites.

A needs assessment undertaken by the UNDP in 2013 around Basrah, Baghdad, and Kurdistan highlighted concern over the impact of war on civilians' living areas and working environments (Zwijnenburg and Wim 2015). As part of this assessment, the UNDP conducted a workshop that brought together representatives from civil society, academia, as well as the private and public sectors to identify the need for stronger environmental regulation, and any other specific needs in that regard. The workshop participants called for the remediation of polluted sites and mines so that the land could be reclaimed and used, and they also lent their general support for cleanup and awareness-raising efforts regarding the environmental issues of Iraq.

That same year, the Iraqi MoE, together with the UNDP and UNEP, drew up a long-term plan that addressed Iraq's main environmental issues. Their National Environmental Strategy and Action Plan for Iraq 2013–2017 (NESAPI) (UNEP 2017) included a strategic analysis of the environmental sector, focusing on both man-made and natural causes of environmental pollution. It outlined strategic objectives across a range of environmental pollution and degradation issues. The NESAPI generated by this collaborative workshop addressed a number of key issues, such as population growth, desertification, urbanization, environmental awareness, and the impact of Iraq's conflicts on its environment. The NESAPI recognized the problems associated with the management of hazardous "conflict waste," seeking to "develop solid and hazardous waste management and assess the pollution of former military manufacturing sites and conflict zones."

The NESAPI is ambitious; based on its strategic objectives, it will require vast amounts of funding, capacity, and expertise to undertake all the assessments, cleanup, storage, and monitoring of both the environment and affected populations in contaminated areas. In January 2014, the Iraqi government, with the support of UNEP and UNDP, published their annual "State of Environment and Outlook Report" (Iraq Ministry of the Environment 2014), which promoted the landmark agreement between the UNEP and the Iraqi Government that "aims to speed up recovery and support peace building." The report noted the impact of war on health and the environment, stating that "Years of conflict and violence resulted in chemical pollution and unexploded ordnances, which is affecting the safety and lives of an estimated 1.6 million Iraqis" (Iraq Ministry of the Environment 2014). Although this number mainly refers to the victims of unexploded ordinances (UXOs), landmines, and cluster munitions, there are serious concerns about the long-term impact of environmental pollution on the general health of Iraqi civilians, from access to clean drinking water, the collapse of waste management, and exposure to a range of chemical pollutants. The legacy of fighting in northern and western Iraq may lead to further breakdown of an already fragile environmental situation, with profound and direct impacts on public and environmental health.

Insecurity and a range of competing political and humanitarian issues may hamper the planned identification, analysis, cleanup, and monitoring of hazardous sites. The situation in Iraq is another example of why the impact of conflicts on the environment requires a collaborative approach, specifically in the form of greater attention from the international community. At present, UN agencies can only act if formally requested to do so by the affected state or mandated to by the UN Security Council, causing delays that can lead to greater impacts on civilian health and increase the complexity of environmental damage. Thus, in addition to efforts to minimize environmental damage in conflict, resolving the legacy of conflict requires greater visibility of environmental incidents and a comprehensive approach to assessment from a more diverse range of actors, such as the military, the de-mining community, international expert organizations such as the WHO and UNEP, and humanitarian organizations operating in the field. A substantial body of information on, and specific expertise in, environmental pollution that is a result of industrial incidents or natural disasters already exists, which could help inform risk assessments and responses to conflict pollution in Iraq. Ensuring that there is sufficient oversight and capacity available to respond swiftly, both during and after conflicts, to reduce the risks of exposure to civilians in Iraq, and indeed in many areas of the globe, is the new challenge for this century. Lessons learned from Iraq and other current conflicts should help inform the development of a set of mechanisms for dealing with the toxic remnants of war.

2.11 The Future of the Iraqi Environment

In additional to the past impacts from war, population growth is currently pressuring Iraq's food, water, and energy resources. By 2030, the population of Iraq is expected to grow to almost 50 million people (Figure 2.9).

The UNEP (2007) found that 5%–8% of Iraq's gross domestic product is lost annually to environmental degradation, and an estimated 39% of Iraq's agricultural land suffered a reduction in arable cropland between 2007 and 2009. Meanwhile, food insecurity is increasing with the quality and quantity of the country's water heavily damaged by upstream damming and pollution,

FIGURE 2.9
(See color insert) The Tigris River, flowing through Baghdad, Iraq. (Photo by Omar Harran.)

and the impacts from climate change and inefficient usage are only making the circumstances worse. As you may recall, around 31% of Iraq's surface is desert, whereas 39% of the country's surface is also estimated to have been affected by desertification, with an additional 54% under threat of desertification. Because of declining soil moisture and lack of vegetative cover, there has been an increase in the frequency of vast dust and sand storms, often originating in western Iraq. Therefore, sustainable access to safe water and sanitation remain as a fundamental challenge. With 83% of Iraq's wastewater left untreated, which contributes to the pollution of Iraq's waterways and general environment, water is destined to be the focal point in Iraq for the foreseeable future. Limited water supplies were reduced still further by a drought that lasted from 2005 to 2009. Over the longer term, the amount of water available per person per year decreased from 5,900 to 2,400 m^3 between 1977 and 2009. "The Tigris and the Euphrates, Iraq's two major surface water sources, may dry up by 2040 if current conditions prevail," warns the government report, "Iraq State of the Environment and Outlook." The report is available (in Arabic only), as prepared by the Government of Iraq, with support from the UNEP, the UNDP, and the WHO (Environment News Service 2014).

To fully understand the current environmental problems in Iraq, and their impact on the people of Iraq, there needs to be a full recognition of the effects of a series of wars over the past 25 years (i.e., Iraq–Iran, the first Gulf War, the U.S. invasion of 2003, and ISIS in 2014) and the subsequent lack of sufficient governmental and international support to sufficiently improve healthy environmental conditions for people, or to restore the ecology of the land to its former functional conditions. With this recognition, it becomes apparent how very useful a few fundamentals could be: (1) an articulated vision, mission, and strategy for environmental policy; (2) international and national capacity building (e.g., training and workshops); and (3) effective implementation of technology transfer, including strategic planning, continuous education, and postgraduate studies, as well as long-term scientific research programs (utilizing international funding donors such as United States Agency for International Development (USAID), European Union (EU), World Bank, and other NGOs). These fundamental factors/resources have great potential to improve the environmental conditions and heal the ecosystems of Iraq.

Assessing these factors on a national scale should help to identify important areas and topics for environmental assessment. For the natural environment, the indicators of importance are biodiversity and sensitivity to degradation. In general, the industrial areas in Iraq are sited in heavily developed regions of low biodiversity and sensitivity. Biodiversity is particularly low in the central plains of Iraq and the Baghdad region due to a history of sustained agricultural activity and the population density in these highly river-exploited areas. However, the most significant environmental receptors (to indicate potential vulnerabilities to people and the environment) are the Tigris and Euphrates river systems, as well as the underground water aquifers of the area. In the south of Iraq, the most important areas to monitor are the Iraq Marshlands, riparian zones, the estuarine Shatt Al-Arab, and the coastal mudflats.

To accomplish the important goals of restoring Iraq's deteriorated environment, the Government of Iraq has signed a landmark cooperation agreement with the UNEP in 2014 to speed environmental recovery and support peace building. This strategic cooperation agreement with the UNEP is intended to strengthen Iraq's attempts to overcome the country's many environmental challenges. Areas of cooperation defined by the agreement are environmental legislation and regulations; biodiversity conservation; the green economy; clean production techniques; resource use efficiencies; combating dust storms; and climate change reporting, mitigation, and adaptation (Environment News Service 2014).

It is important to point out that the Government of Iraq has made definitive progress in building a new government, approving a constitution to supersede the Saddam Hussein era, and holding successive elections for parliament and provincial governments. However, governing institutions remain weak, and corruption and poverty are pervasive. In February 2018, an international conference opened in Kuwait to plan ways to rebuild Iraq and secure it against renewed extremist violence following the 3-year war against ISIS. A U.S. Institute of Peace (USIP) team also recently spent 9 days in Iraq for talks with governmental and civil society leaders, part of a multiyear-long effort to help the country stabilize. The USIP conference (in Kuwait) gathered government, business, and civil society leaders to consider a reconstruction plan that was estimated by Iraq to cost $100 billion. The USIP takes the position that any realistic rebuilding plan must focus on the divisions and grievances in Iraq that led to ISIS' violence. It is evident that Iraq will continue to face the many challenges to rebuild their formerly healthy environment and substantial impediments to reinvigorating national infrastructure, which will likely take decades to overcome. In the interim, the approach being implemented is dominantly driven by a collaborative method, utilizing an international consortium of nations, which utilizes the experienced coordination of the United Nations and others.

Following the complete liberation of all Iraqi territories from ISIS in December 2017, the Government of Iraq put in place a comprehensive reconstruction package linking immediate stabilization to a long-term vision, and initiated a recovery and reconstruction process. The ISIS war and a protracted reduction in oil prices have resulted in a 21.6% contraction of the non-oil economy in Iraq since 2014, contributing to a sharp deterioration of fiscal conditions in the country. An improvement in the economic factors, namely, higher oil prices, as well as better security in 2017 contributed to an improvement in economic stability for the country and a return to growth in the non-oil sector.

With the improvement in the economic fundamentals of Iraq, which ironically depends on the success of the country's oil industry, the hope and the intent of the partners in Iraq are that, now, Iraq's environment and its sustainable resources can begin the steady path to recovery, which will ultimately lead to improved and sustainable public health, lands, water, and agricultural systems.

After such a dreadful story of conflict, and restoration, and with a glimmer of hope for the restoration of rich traditions and the land of Iraq, let us now journey to the other side of the world to experience the close ties between a watershed and the human spirit. Exploring the undulating landscape of the Ozark Mountains, what follows in Chapter 3 is a story of collaboration and transformation driven by the close ties of the people of the Upper White River Basin with their water source, reminiscent of the Ma'dan's closeness to the rich marshlands of southern Iraq.

3

Water Quality as a Collaborative Force in the Ozark Plateau, Missouri and Arkansas: The Long-Term Dimensions of Action through Common Interest

Shawn Grindstaff
Facilitator and Process Consultant

Brenda Groskinsky
Environmental Scientist

Maliha S. Nash
U.S. Environmental Protection Agency

Ricardo D. Lopez
USDA - Forest Service, Pacific Southwest Research Station

CONTENTS

3.1 Introduction

Water is a primeval element of human existence. At its most basic, water is life. Water is both symbol and shape, fluid, and concrete. Water is connected to time, by the second, and yet there is something timeless about water as well.

People cannot resist the draw of water. We are born in it, sustained by it, refreshed and rejuvenated by it. We attempt to use it, and even attempt to control it. At many times, we are overpowered or forever altered by it. Water challenges our very notions of how people receive information, how they process knowledge, and how they feel. Everyone knows water, and yearns for their own story that weaves it into the tapestry of their life's journeys. In the study and work of water, we traditionally engage it in the context of policy, government, engineering, and other scientific and technical disciplines. Those endeavors, although important, rarely take us to the depths of human celebration and camaraderie necessary to create the place and space for spiritual purpose, and practical action about water. The emotional elements of water beg for more attention to it, in policy and management circles, and its management in the future. Its familiar tug on us often leads us to a place of emotion, and at times a place of celebration, where success, purpose, and family are just as important as the science surrounding the water itself.

Given this tight relationship between physical water and our lives as humans, every life and every river has a story to tell. As such, we share one such story, a story of the past that also initiates hopes for the future. And like the stories told since the dawn of humankind, we draw ourselves in and we listen, looking for that which we seek in our own lives. Such is the story from a culture of proud storytellers—the people of the Ozark Plateau, who have always seemed to know that the future points just beyond the river's bend.

This is a story about the people with their lives interwoven with the landscape of the Upper White River Basin (UWRB). It is a story about people and water; a people who are filled with pride, a strong work ethic, and resourcefulness. It is a unique culture in the lower Midwestern United States, where an ancient plateau has eroded itself under the elements of time, sculpting some of the world's most beautiful forested hills, hollows, and glades across a significant expanse of southern Missouri and northern Arkansas. A place, much like the lives built on them, where the hills stand out in inspiration and the valleys cut deep, as the water falls, and then again rises, with the urgency of the communities it sustains across the expanse.

The Upper White story is about people steadfastly protecting their water over a period of great change—a common theme that spans all of time and life, across the world. It is as if the people of the UWRB are playing instruments in an orchestra, at a watershed scale. They are on stage, individually and collectively playing an epic and memorable composition, full of movements and immersed in all of the elements of music. Every person, from all walks of life in the Basin, is performing their part for a moment, destined for the millennia. Each musical element represents the opportunity to cherish, remember, and learn. The efforts that they exert are the celebrations that are part of a continuum, in recognition that what people have is good and that the most valued assets of their lives, what gives it meaning, bridging past, present, and future is the watery landscape itself. It is a feeling of a place, water's life-giving place among it, and the emotions it can evoke. It is a place people call "home" and a home they call their "place," among the hills. This is the story of the western edge of the Ozark Plateau, home to the large watershed known as the UWRB.

3.2 Hollows and Hip Waders[1]: An Analysis of How Success Is Formed

As in so many environmental events in history, there seems to be a pivotal moment that provokes action to fix a problem, what we have referred to earlier as a societal imperative. Often it can be a devastating event or public health emergency, accompanied by government intervention and problem-solving teams of the public sector, academia, and nongovernmental organizations focusing from a negative start point with a constant "fix" and "solve" perspective. But what happens when the pivotal event does not occur in that

---

[1] "Hollows" and "hip waders" are two common Ozark terms. Hollows (often pronounced "hollers" in the Ozarks) denote small, often deep valleys where streams and rivers have carved out sheltered areas suitable to hunt, fish, farm, or live. Hip waders are waterproof boots extensively used by people who fish and wade the Ozark streams and rivers.

way, and small, seemingly disconnected incremental moments of awareness diffuse across an entire population of a watershed? In other words: what inspired people to initially act, and continue to act, and to evolve and still aspire to protect their water in a way that will benefit generations to come?

To discern this spirit of a people who made hundreds of sometimes connected, sometimes disconnected, strides in protecting their watershed, the typical problem-solving models cannot answer what motivates and inspires normal citizens to do extraordinary things on behalf of their water and their communities. This quest for the root causes of success leads one to a different set of questions to frame and unveil this unique feat. In such a paradigm shift, a natural question would be to research what went right and what things can be celebrated, either as stand-alone moments or things that can be replicated from others' wisdom and toil.

With this goal in mind, this chapter focuses on the authors' own unique vantage points of the entire watershed, revisiting inspiring cultural stories of all time on a watershed scale. Resourceful and amazing people, in their own words, reflect and recount inspiring moments, events, and people that catalyzed awareness and action to protect their water. Written through the perspectives of on-the-ground history, this brief introduction to modern Ozark water history is generated from and spoken by those who were there, "in the hollows and hip waders," which has truly transformed the trajectory of Ozark water quality awareness and action.

3.3 The Geography

In the minds of those Ozark storytellers, the tale of the UWRB must begin with a basic overview of the watershed as if the reader were a drop of water making your way up, down, and around the rich landscapes and through the scores of rural villages and cities along the way. The White River, constantly fed by its tributaries, big and small, takes an unusual and counterintuitive path from its headwaters (like a comma turned sideways) to its watershed endpoint, where the White River leaves the watershed on its way to the southeast (Figure 3.1). The unusual aspect of this snakelike journey is that it begins in the majestic Boston Mountains of northwest Arkansas (the highest points of the Ozark Plateau) and veers north into Missouri before returning south into Arkansas and then to the southeast through the state. The White River's 722-mi (1,162-km) journey from birth to passing flows through a total of eight dams, a namesake National Wildlife Refuge, and a multitude of Ozark cultural landscapes on its way to the cotton and rice of Arkansas' Delta region, and then to its confluence with the Mississippi River.

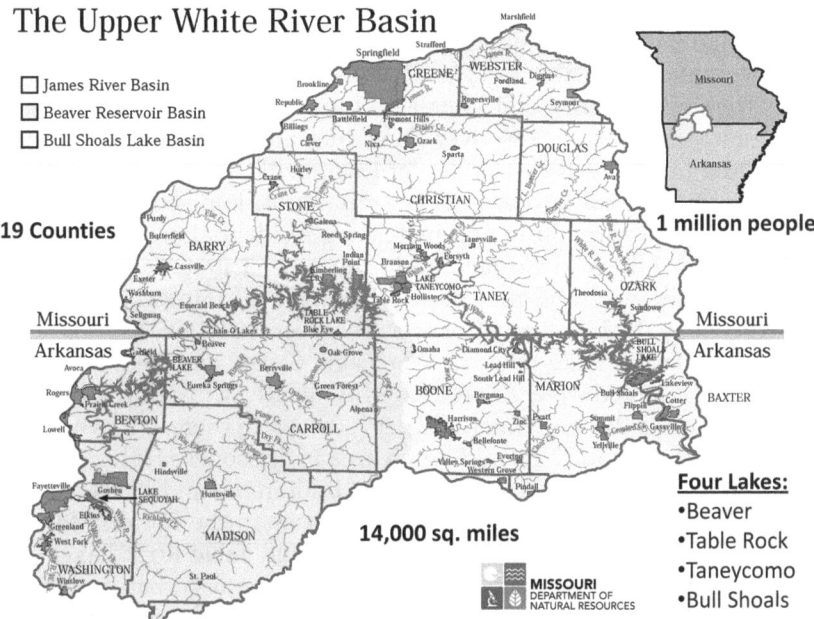

FIGURE 3.1
(See color insert) The UWRB lies within both states of Arkansas and Missouri and flows over Ozark topography,[2] which contains rural agricultural and residential areas, multiple man-made reservoirs, and ultimately flows into the Mississippi River.

3.4 If We Were Water: A Trip down the Upper White River[2]

The trek begins in the White River headwaters sub-watershed, at the source, approximately at a Boston Mountains farm pond at an elevation of 2,250 feet above the sea level (Beaver Water District 2008). As is the case in the Ozarks, the collecting rainwater goes to gullies and small brooks through vibrant oak/hickory forests heading down through winding creeks until a river forms below. This transformation happens near Arkansas Highway 16 as other creeks and small tributaries support the "birth" of the White River. As the river winds through small cities such as St. Paul, the landscape changes from the remote mountain forests to the valley-area agricultural look and

---

[2] "A fairly important characteristic of the Ozarks… is the fact that a large proportion of water flow in the region is underground" (Lopez and Frohn 2018). The Ozarks' topography is a landscape portrayed by soluble rock formations that have been dissolved over time by both surface and groundwater. This "karst" topography is unique and only "exists across approximately 10% of the Earth's surface, and 25% of water relied upon by humans worldwide." "Understanding the conditions of karst in the Ozarks… is important because of the relevance to determining the vulnerability of water resources to water pollution" (Lopez and Frohn 2018).

feel, as the water nears better land for farming and supporting agricultural pursuits like cattle and the significant industries surrounding poultry. As the river then enters the Lake Sequoyah–White River Sub-watershed, the water moves from mixed forest and some agricultural lands to a rapid increase in new residential areas, as the river flows into the 490-acre Lake Sequoyah, constructed in the 1960s by the City of Fayetteville, Arkansas. Just past this point, the Middle Fork and West Fork of the White join, and it moves north. The mix of rural agriculture and increasing residential areas characterizes the path from Elkins to Goshen, taking in other tributaries like Richland Creek along the way, and then slowing to the beginnings of Beaver Lake. Constructed by the U.S. Army Corps of Engineers (USACE) from 1959 to the early 1960s, Beaver Lake takes in War Eagle Creek, and then Beaver Dam discharges the stored water to a reformed White River, just downstream near Beaver, Arkansas (Beaver Water District 2008). From there, it moves through the Chain-O-Lakes and Emerald Beach area in Missouri as it becomes Table Rock Lake, another USACE impoundment built in the mid-1950s. The Kings River, another tributary of the White that forms in the Boston Mountains, travels north about 90 mi (145 km) to empty into Table Rock Lake. On the north end of the impoundment, the James River, the most urbanized tributary in the watershed enters from the north. Below the dam, the eastbound river passes through Branson, Missouri as Lake Taneycomo, passing through the 1913 utility-built Powersite Dam at Forsyth, Missouri, then veers southeast again and through the USACE built Bull Shoals Lake across the Arkansas state line (Figure 3.2). That dam was constructed in the late 1940s and early 1950s. At that point, the river winds to the southeast, past Flippin and Cotter, Arkansas, and out of the watershed as it descends finally to the Mississippi River lowlands in the Lower White River watershed. The White River then enters the Mississippi River in Desha County, Arkansas, only miles from the

FIGURE 3.2
The Upper White River, just below Bull Shoals Lake. (Photo: Ric Lopez.)

entry point of the larger Arkansas River, and about 100 mi (161 km) south of Memphis, Tennessee. All told, the UWRB covers 19 counties in Arkansas and Missouri, spans 14,000 mi² (36,260 km²), and is home to approximately 1 million people (Ozarks Water Watch 2018).

3.5 History Spoken by Those Who Shaped It: Inspired Hope for Generations to Come

From the first written accounts by Henry Rowe Schoolcraft in 1818–1819 about the Ozark landscape (Schoolcraft 1821; Schoolcraft 1996) to the more contemporary understandings of the Ozark's ties to land and water (Campbell 2010), it is clear that rocky, vertical landscapes, and karst geology typical of the UWRB had a pivotal role in the cultural development of the area. The lack of human-built infrastructure across the remote expanses of the watershed until the mid-20th century was a significant factor in the culture, on most levels. The relative lack of large-scale agricultural capability also played a role in holding back a wide variety of livelihoods and land uses so prevalent in other areas of the country, while building generations of strength and character in the people who settled and remained in the watershed.

Change arrived though, in the post-World War II era, as the push to develop and advance progress began to make its mark in the watershed. There would be no bigger display of this push than in the study, construction, and operation of dams up and down the White River. Much like other American rivers that levied destructive flooding that took life and property, the Upper White River became the focus of a series of flood control dams (among other purposes) that could attempt to bring both relief and promise to the watershed. This mammoth federal government civil engineering feat from 1947 to 1961 changed the river and the entire watershed forever. The federally funded reservoirs were created for the core purposes of flood control, hydropower generation, and water supply in that historical era.[3] However, as the environmental era began in earnest, post-1970, other purposes were added to these massive impoundments. In Arkansas, e.g., the state added recreation and aquatic life as a designated use to Beaver Lake under the Clean Water Act (Beaver Water District 2008). This cluster of uses helped trigger unprecedented development, economic activity, tourism, and recreation as a value system and business sector in and of itself. The region then absorbed all the corollaries one might imagine, from aesthetic population influx, retirement communities, and scores of interconnected ventures, all tied

---

[3] "The growing demand for irrigation, water supply, and hydroelectric power as population and economic growth surged after the Second World War led to an unprecedented boom in dam and reservoir construction…" (Postel and Richter 2003).

to the Upper White water. The three new dams and the lakes that resulted in the Upper White set the stage for the common interest and economic drivers that would create an amazing movement to protect water quality. Interestingly enough, Postel and Richter (2003) commented that "Most rivers are no longer controlled by nature, but by us."[4] If this is true in the Upper White River in the Ozarks, then it may have been that gradual handoff of control that presented the birth of greatness and a call to action for citizens in this new landscape, and reality developing around them. Indeed, the water and the culture, tied together in spirit, had to redefine their relationship.

When singular events or tragedies do not shock people into response and action, the only alternative is to draw from some combination of other motivations and dare to build a system from the ground up. This system must be adaptive while enabling people to find their own way to protect and cherish their water. It will occur because they are using their own sense of place and perspective, their own dreams, and their own resources. It is an organic approach well suited to thrive in the Ozarks.

In other words, if moments were stars, it takes a constellation of "ahas" to form and advance stewardship momentum at a watershed scale. In the UWRB, it happened.

3.6 In Their Own Words: Discernment and Documentation of Inspiration

For purposes of celebration, a series of questions were framed for many of the most astounding Basin individuals and champions of water stewardship over the period of 1993–2018. Their one-of-a-kind stories, journeys, and wisdom were gathered throughout a year of conversations (2017–2018) to create a true perspective of the magnitude of human accomplishment and their humility amidst the thousands of others whose offerings could not be covered in this short chapter. Their quoted comments are attributed to them as a group and are anonymized to illustrate their collective wisdom and patterns of common experience and musings, about the societal dimensions of their "water story."

3.6.1 "What Inspired You to Get Involved or Take an Interest in the Quality of Water in the Upper White?"

For most of the people interviewed, they spoke with great love for the abundant water resources in the UWRB, and the beauty and awe of the

---

[4] Water flow modification/engineering plans by humans can be dated all the way back to the 16th century and a collaborative effort between Leonardo da Vinci and Niccolò Machiavelli to modify the flow of the Arno River (Masters 1998).

Ozark landscape. Coming from very different walks of life, it was clear that so many of them had always enjoyed the resource in their own ways in their early years. However, they mentioned that over time, the population growth, economic development of the bistate region, and other pressures created feelings ranging from "nervousness," to "a slow awakening," to the challenges that were mounting to key rivers like the James River in Missouri and the Kings River in Arkansas, not to mention the White and its four large lake impoundments. For some, there was a business interest in keeping the waters clean. With a burgeoning tourism industry in places such as Branson, Missouri, and Eureka Springs, Arkansas, a business interest in clean water was relatively obvious as challenges arose in the late 1990s due to population growth across the region. Another person noted that due to the mild American southern climate, the very same areas were becoming popular zones for retirement communities and associated development. Still others spoke of their love for the large lakes and that as the sport fishing and assorted recreational sectors increased in popularity, there was also a keen interest to keep the waters healthy both from an environmental and aesthetic standpoint. For those raised in agricultural areas, the journey was different but still keyed on finding the best way to be good stewards of both the water and the land they owned and loved. In their worlds, they spoke of two distinct eras in their lifetimes. In the early years, the focus growing up was on controlling erosion, a substantial and common occurrence in the rolling topography of the watershed. With the inherent difficulties of farming in the Ozarks due to topography and soil quality, this continues to be of interest to many and always will be a stewardship focus. A later focus for them was how to responsibly realize new economic opportunities in agriculture and food production with crops, cattle, and poultry production. As one person articulated, "putting food on the table for my family by putting food on the table for other families."

What is notable is that many of the people of the UWRB talked about the late 1990s as the period when many of these independent events began to gain interest in various civic and small group conversations. There was also excitement due to the rapid growth and economic opportunity across an Ozark watershed that, according to everyone, had never happened in the history of the UWRB. The stage was being set for a focus on water quality, but it was during a corresponding era of tremendous excitement and a new quality of life that generations had yearned for in the Ozarks. Among those who remember, an individual noted that "it was the building of the exciting infrastructure, the new stores and new highways and new businesses itself that served as a laboratory of what happens to karst geology under pressure, not to mention the difference between impervious surfaces and what was there before that. It was exciting, but we were in a learning mode during all these years."

A first "aha" moment springboard for ideas apparently started with a group of interested citizens in 1984 in Springfield, Missouri (a large city at the north end of the watershed), that met monthly and has continued that tradition ever since. Those aware of these early gatherings acknowledged that it has been a remarkable feat that that dialog has continued unabated for 35 years, creating the impetus for development of the Watershed Committee of the Ozarks. The Committee is one of the several prominent watershed groups that operated, mostly in Missouri in that era.

Over the past 25 years, many watershed organizations have flourished in the Missouri portion of the watershed, and numerous smaller organizations also operated in the Arkansas portion of the Upper White. For better context, we offer the Internet websites to garner more information about their history and missions. These key organizations, provided below in order of their original founding date, helped develop context for the wisdom of the people who contributed pieces of their journeys to this story. It is by no means a complete list below, and numerous other organizations and individuals played their roles and are referenced in the websites listed below as well. Note that there are so many contributors to the work in the Ozarks that, due to length limitations of a written book chapter, we are unable to fully expand on their contributions:

- Beaver Water District (Arkansas)—founded in 1959 and is a public water supplier in Arkansas. See www.bwdh2o.org (checked September 15, 2018)
- Watershed Committee of the Ozarks (Missouri)—founded in 1984. See www.watershedcommittee.org (checked September 15, 2018)
- James River Basin Partnership (Missouri)—founded in 1997. See www.jamesriverbasin.com (checked September 15, 2018)
- Table Rock Lake Water Quality, Inc. (Missouri)—founded in 1998 and merged into Ozarks Water Watch (OWW) in 2014. See www.ozarkswaterwatch.org (checked September 15, 2018)
- Upper White River Basin Foundation (Missouri)—founded in 2002 and now is known as OWW. See www.ozarkswaterwatch.org (checked September 15, 2018)
- Beaver Watershed Alliance (Arkansas)—founded in 2011. See www.beaverwatershedalliance.org (checked September 15, 2018)

3.6.2 "What Happened When Science Was Included in the Developing Conversations?"

In their own words:

> The science of water really kicked in and was useful from the mid-1990s on, kind of like the government and their involvement in general. Our conversations started first, they came later…

Environmental science gave us the tools to address what we were seeing but we didn't fully understand back then…

The work of the scientists at OEWRI [Ozarks Environmental and Water Resources Institute, College of Natural and Applied Sciences at Missouri State University] and AWRC [Arkansas Water Resources Center] was immensely helpful to frame up what was happening. The Missouri Department of Natural Resources and U.S. EPA were both very helpful and appreciated.

The work of Arkansas agencies like the ADEQ [Arkansas Department of Environmental Quality] and ANRC [Arkansas Natural Resources Commission] were essential to help us move ahead.

3.6.3 "What Were the Three Greatest Milestones of the Past 25 Years in Protecting the Waters of the Basin?"

In their own words:

The major algal blooms of 1997–1999 on the James River near Table Rock Lake made an impression on a group of citizens enough that the James River Basin Partnership was formed.

When the City of Springfield voters overwhelmingly approved a bond to help the Southwest Treatment Plant help reduce phosphorus discharge into the James, while others like Branson, Ozark and, Nixa did their part too through a period of time.

The four major non-profits were formed or ramped up during this time period.

Awareness was born during this last 20 years and it will now last for generations to come. It came from many and not one place. But it is here now.

Using water seems to be cast in often negative terms in other places because of conservation concerns. However, in the Ozarks, this is a positive because you can't protect it if you don't use it and own the benefit of it. Usage equates to awareness. Awareness leads to a better place.

A recent grant represented the first true joint application of several key groups, allowing a newer, more direct way to leverage each other from the application stage onward. Ways to improve and adapt to constantly changing environment is getting more and more refined. Years of challenge makes you better at meeting it.

At some point, there came a recognition that we have challenges without intention and awareness, and that positive action was required. The other recognition was that the status quo wasn't working and that we can't keep doing things the same when things affecting the water are constantly changing around us.

The Beaver watershed convening process in 2009 and 2010, leading to the Beaver Watershed Alliance from 2011 onward. It was a big thing and led to real source water protection efforts.

The Source Water Protection plan for Beaver Lake, one of the first in the nation, and the 4 cents per 1000 gallons (3785 Liters) sold at the Beaver Water District to help sustain source water protection efforts over time.

3.6.4 "What Were Your Three Greatest Personal Moments in the Quest to Improve Water Quality in the Basin?"

In their own words:

When I was out canoeing the James River this last summer and saw in vivid detail how healthy it was that day. Every detail and past thing had culminated in that one moment for me.

The day it became clear that the Northwest Arkansas Council was going to focus on the Beaver Lake watershed and that source water protection may actually happen.

The day the Board of the Beaver Water District voted to dedicate 4 cents per 1000 gallons (3785 Liters) to source water protection programs to help sustain programs and efforts to protect the Beaver Watershed of the Upper White.

The minute I realized that many people acknowledged the 17-mile [27-kilometer] algae bloom in 1998 and that they were actually going to step up.

In the first phase of tackling point source pollution, the City of Springfield's courage and common sense.

Every day another group or effort found funds to sustain them another month or a year to help the water.

The week I attended a national watershed conference back in the early 1990s. It woke me up. My whole world opened up for me. I found I wasn't alone in my walk, and that there are always new ideas.

The culmination of so much in the formation of Ozarks Water Watch.

The day I realized that so much good had happened in the Upper White that similar efforts began to take root and develop in adjoining watersheds like the Illinois River in Arkansas and Oklahoma.

The day in 2008 the two governors of Missouri and Arkansas signed the Bi-State MOA in Springfield, Missouri. Through the efforts of Ozarks Water Watch and so many others leading up to that moment. So many others I can't remember them all.

3.6.5 "What are the Three Qualities of the People of the UWRB in Both States that Deserve Celebration?"

In their own words:

The joy of seeing citizen engagement in their water nowadays, like River Rescue, the lake-cleanup events or Secchi Day on Beaver Lake.

County-wide education festivals on water.

Catalyzing young people to pursue their dreams to help protect water in their career goals.

People here have an appreciation of water itself, and now appreciate the karst geology and uniqueness of it all together.

The people here are of the land. They have a cultural stewardship and are protective of their water.

A group of people have emerged who have become committed to streams going to the lakes, and they focus on heightened awareness, the quality of feed streams, and stream bank repair.

People of the Ozarks won't be cow-towed or coerced when it comes to their water—it is their own self-interest. They are independent thinkers, and that is an asset!

People here appreciate the uniqueness of where they live, and they have a realization of it now.

Ozarks people have a sense of humor and it is "self-deprecating." They don't take themselves too seriously while they work.

Humility. People are humbled by the water.

The culture and history of Ozark people and their water is centuries old.

People here are to be celebrated because they are individual, independent, outdoor in character.

Whether it's a farmer doing best management practices or other good people doing good things, there is a selflessness here to do good for generations to come. To be stewards of God's blessing.

3.6.6 What Have We Learned from Our Successes and the Wisdom of the Decades?

In their own words:

We are still finding ways, after all of these years, to find new ways to work together in our common interest.

Everyone agrees education is important, but the question is always how to prove it is happening, how you do it, and if it is effective at all on its own.

In the Ozarks, and maybe everywhere, there is no substitute for face-to-face communication. It will be the most important way forever.

We all have a part in the contamination of the watershed and we all ought to be the answer.

Everyone has facts. But the key to having knowledge is understanding the interrelationship of everything. To do that, you must build relationships first, and have technical knowledge second. They both have to happen, but in that order to really work at the watershed level.

Watershed work of any kind is hard because it is never still. The key is perseverance.

A word to the next generations... if you want to do the right thing on protecting watersheds and water quality without cutting in on other's well-being communicate, collaborate, get involved. That's what has to happen to have lasting success.

3.7 Baby Steps and Boldness: "Aha" Ideas Become Innovations (2002–2018)

The following efforts emerged from all of the energies that surrounded the Ozarks, and their initial steps have taken bloom, becoming the innovations that propel the region into the future:

1. Four watershed groups (Watershed Committee of the Ozarks, James River Basin Partnership, Table Rock Lake Water Quality, Inc., and the Upper White River Basin Foundation) worked on a series of meetings and watershed summits from 2002 to 2003 to have dialogs about water quality across the Missouri portion of the UWRB. The UWRB, the only one of the four with reach across the whole Basin, convened a second Summit in March 2005 covering the Arkansas portion of the Basin with a group of Northwest Arkansas-based nonprofits. In November 2005, UWRB convened the third and final Summit over the entire Basin, the first bistate watershed-wide consensus-building dialog. More information is available from OWW at www.ozarkswaterwatch.com (checked September 15, 2018).

2. The "Bi-State Memorandum of Agreement Regarding Cooperation on Water Quality and Water Quantity Issues in the States' Shared Water Resources" of November 24, 2008, signed by then Missouri Governor Matt Blunt and then Arkansas Governor Mike Beebe pledging that the two states will enhance and promote cooperation among the state agencies, address both surface and groundwater resource issues, acknowledge that their shared water is economically and environmentally important, and formalize the two states' intent to cooperate in addressing water issues of common concern. More of this information is available from the OWW Internet site: www.ozarkswaterwatch.org (checked September 15, 2018).

3. The Beaver Watershed Alliance—After an effort that produced a Lake Watershed Protection Strategy in 2009, this process led to the formation of the Beaver Watershed Alliance in 2011, and they adopted the Lake Watershed Protection Strategy document. More information can be found on www.beaverwatershedalliance.org (checked September 15, 2018).

4. An innovative Source Water Protection Plan adopted by the Beaver Water District Board in 2012. More information and a full copy of the visionary plan is available at: www.bwdh2o.org (checked September 15, 2018).

5. The Source Water Protection Fund vote of the Beaver Water District Board—On April 21, 2016, the Board approved a motion to dedicate

$0.04 per 1,000 gal. (3,785 L) of water sold to the Source Water Protection Fund. Their Vision for Source Water Protection states that "Providing high quality drinking water starts with protecting the source of that water, which is the first step in the multiple-barrier approach. Beaver Water District will lead the citizens, businesses, and communities of Northwest Arkansas to cooperatively maintain the quality of Beaver Lake for all generations."

6. The Upper White River Basin Foundation, now doing business as OWW, is a water quality organization that has focused on preventing pollution through improvements in wastewater treatment and septic systems in their area of the Ozarks. Currently, OWW is housed with two other sister 501(c)(3) nonprofit entities located in Kimberling City, Missouri, with an additional office in Rogers, Arkansas. Ozarks Environmental Service specializes in wastewater maintenance of over 70 wastewater treatment plants, whereas Ozarks Clean Water Company owns 18 wastewater treatment systems. The combined mission of the three organizations is, "To protect water resources and the public health by keeping our Ozarks waters clean and clear." Mergers over the years with other organizations (i.e., Table Rock Lake Water Quality, White River Environmental Services, and English Village Nonprofit Sewer Company) have brought additional expertise and resources to the group. More information is available from the OWW Internet site: www.ozarkswaterwatch.org (checked September 15, 2018).

7. In 2004, The Upper White River Basin Foundation requested assistance from U.S. Environmental Protection Agency (EPA) Region 7 in the development of a methodology that would help the UWRB prioritize water quality improvement efforts in the Upper White River watershed. As a result, U.S. EPA Region 7 initiated a collaborative research effort with U.S. EPA's Office of Research and Development to map and interpret landscape-scale ecological metrics among the watersheds of the Upper White River. The outputs of the project were specifically designed to address stakeholders' concerns about water quality by optimizing the application of ecological restoration efforts throughout the Upper White watershed.

The U.S. EPA assessment of the cumulative impacts of land cover alteration on surface water quality for the UWRB in southwestern Missouri and northwestern Arkansas was initiated for, and in support of, local watershed planning groups and their activities. A mapping and interpretation of landscape-scale ecological metrics within the Upper White watersheds was performed and resulted in the first set of geospatial models of water quality vulnerability in the Ozarks. The models were developed using existing field water quality monitoring data, remotely sensed National

Aeronautics and Space Agency satellite imagery and other remote sensing data, and the theoretical relationships between landscape conditions and the water quality of streams and rivers in the associated watershed(s).

The resulting outputs of information (provided to the people of the Ozarks and the U.S. EPA's Regional Office in the area) was deduced using partial least squares (PLS) analyses, a statistical approach described in additional detail in Lopez et al. (2008). The Ozarks study area consisted of 244 separate sub-watersheds, with each sub-watershed represented by an area of surface water runoff with a single hydrologic outlet point, known as a "pour point." Stream total phosphorus concentration, total ammonia concentration, and total *Escherichia coli* (*E. coli*) cell counts were extracted from monitoring databases. Water quality measurements were compared with 46 broadscale landscape metrics (e.g., percent forest, urban, and agriculture) using PLS to assess how sub-watershed condition correlates with water quality at pour points. Model trends among sub-watersheds were verified with an analysis of four progressively finer scales ranging from the entire sub-watersheds to directly adjacent to the stream bank. PLS geospatial statistical analysis is different than standard regression because it accounts for small sample sizes, missing data values among sampled areas, a large number of predictor variables, correlated predictors, and high noise-to-signal relationships. Notably, all of these PLS characterizations reflect the data sets that are available for the Ozarks. The PLS model was built on 10 non-nested watersheds and then used to predict the total ammonia, total phosphorus, and *E. coli* for the 244 sub-watersheds (Figures 3.3–3.6). The amount of variability in total phosphorous, total nitrogen, and *E. coli* counts is explained by each PLS model reflecting the composition of the contributing landscape among the watersheds analyzed.

The predicted surface water landscape indicators were associated with measurable error values so that decisions about where to perform restorative actions would be known. The error value predictions were determined using all available data, without a probabilistic sample design, and thus increased the accuracy of water quality vulnerability predictions within a known error range. The results provided watershed managers with the first broadscale predictions that could be used to explain how land cover type, land cover configuration, environmental gradients, and human activities might affect the characteristics of surface water in the Upper White River region (Figure 3.1). Additional information can be found in the works of Nash and Lopez (2010), Lopez et al. (2008), and Lopez et al. (2006).

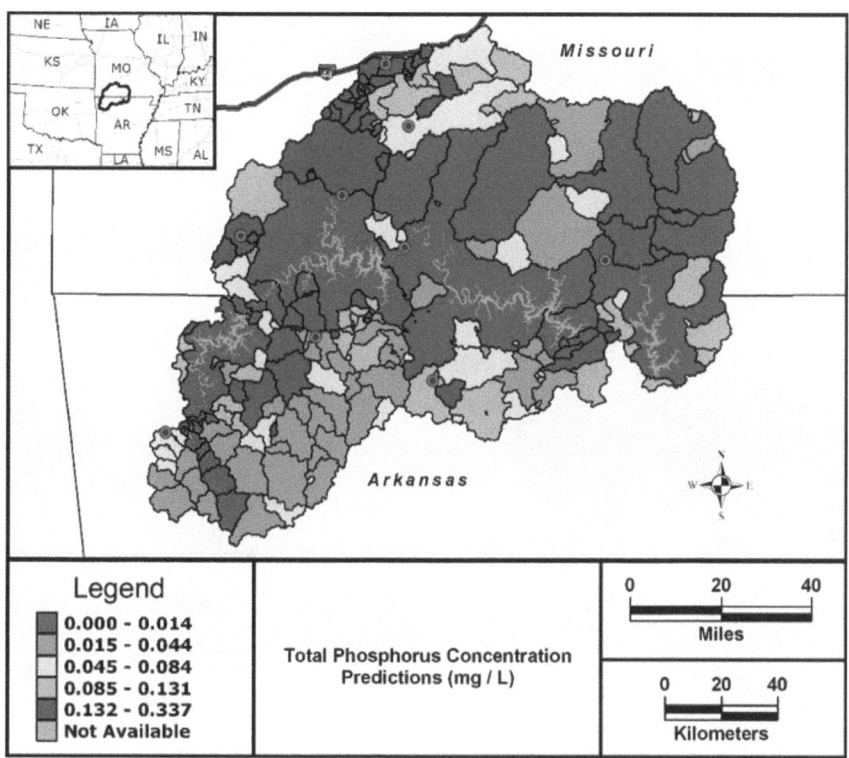

FIGURE 3.3
(**See color insert**) Total phosphorus concentrations among the 244 sub-watersheds in the UWRB of the Ozarks (Lopez et al. 2008).

3.8 Conclusions, Confluences, and Celebrations

It has been acutely apparent to the authors during the development of this chapter that, often, there is a deficit in the amount of reflection and recognition, and associated documentation, of what went right during the societal interactions that occurred when addressing a natural resource challenge such as was faced in the Ozarks. The richness of the journey and the passion and wisdom of so many lives operating in tandem on a bistate watershed such as in the Ozarks has been, and continues to be, an inspiration to us all, who were privileged enough to see the constellation of "aha" moments, up close and among hundreds of good people throughout the several decades of development.

As we stare down the White River, past Cotter, Arkansas, it has become apparent that, although the physical confluence is the Lower White River,

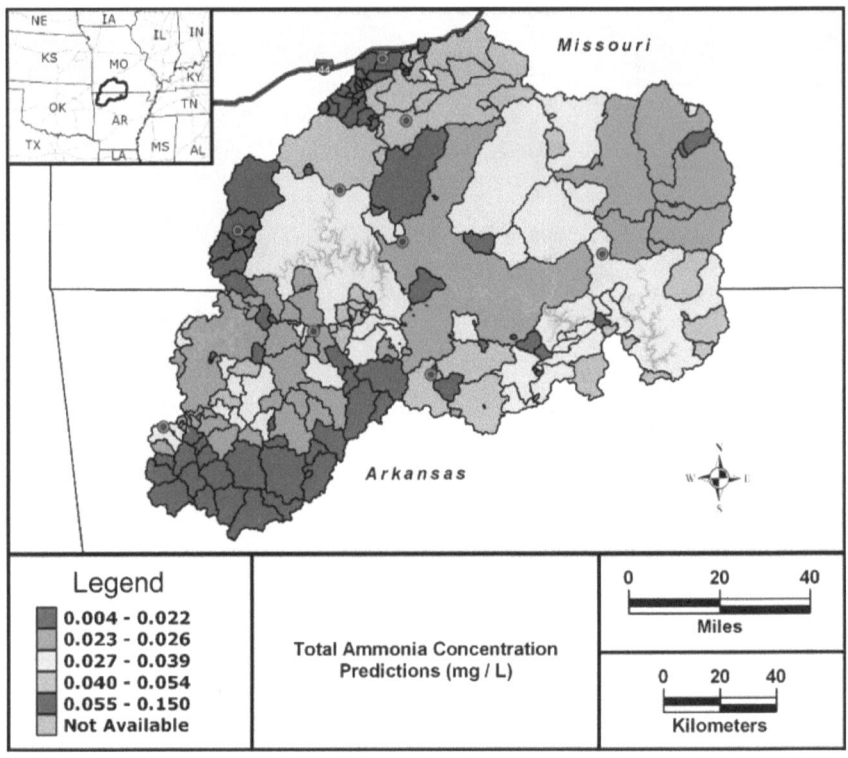

FIGURE 3.4
(See color insert) Total ammonia concentrations among the 244 sub-watersheds in the UWRB of the Ozarks (Lopez et al. 2008).

the metaphoric confluence is the river's visceral importance to key parts of the local culture, the very thing that created the long-term sustainability of a collaborative framework. This framework, the self-professed "cultural stewardship" concept, integrates the cultural and economic and aesthetic fabric of the communities into the environmental science that informs and buttresses their definitive actions. The stewardship concept in this area of the world related the common interest in water, much like a physical tributary would feed a river. Ozark culture, business, tourism, and interest in human health and environment also contributed, all flowing into a large "river" of watershed protection, lending a new clarity to the complexity of the landscape and the necessity of addressing this complexity. This metaphoric river system suggests that policy features a circle of clasped hands, with those of the Ozarks encircling our common interests, but on some magical level of emotional investment, adaptation, and reinforcement of succession planning. It forges a path to wisdom of the future in this wondrous place, the Ozarks. Yet, there are many miles to go before the river system is whole,

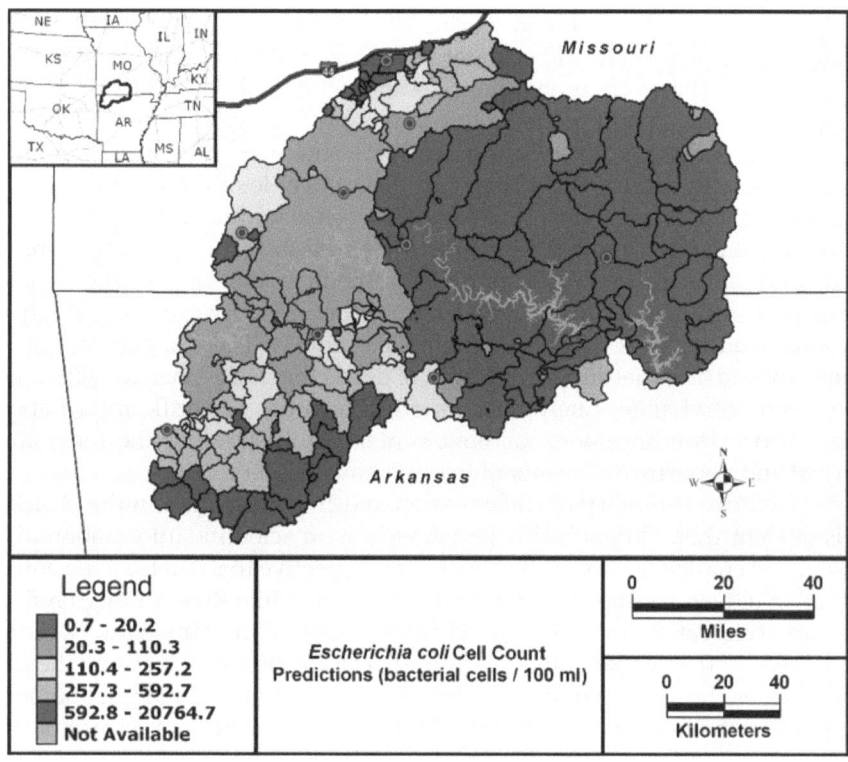

FIGURE 3.5
(**See color insert**) Total *E. coli* concentrations among the 244 sub-watersheds in the UWRB of the Ozarks (Lopez et al. 2008).

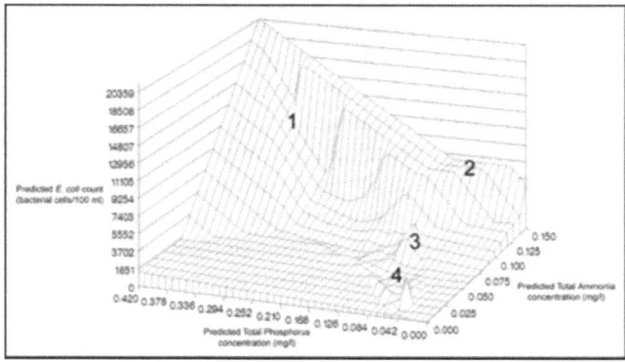

FIGURE 3.6
Three-dimensional plot of predicted total phosphorus concentrations, total ammonia concentrations, and *E. coli* counts among 244 sub-watersheds in the Upper White River region of the Ozarks. Graphic integration of the predicted values generally depicts (1) most vulnerable sub-watersheds, (2) highly vulnerable sub-watersheds, (3) moderately vulnerable sub-watersheds, and (4) least vulnerable sub-watersheds (Lopez et al. 2008).

and the future will undoubtedly bring new twists and turns for the UWRB Watershed. The quest for clean water will go on, mysteries and unveilings at every turn of the river among the trees and fields, just out of view. And the people of the Basin will predictably and reliably rise again to the occasion.

We know of this predictability because of the manner and sequence of their awakening and action. The strong and proud people of the Ozarks culture propelled the resolve, the culture elevated a sense within each of those who devoted themselves to the river, rising to the occasion. The common interests and common landscape of the entirety of the people of the Ozarks created an opportunity for common sense and that common sense, in turn, stitched together a quilt that is the developing policies that will last for generations to come. Indeed, it is that supportive quilt of deeds that is destined to withstand the winds of change, likely much more than pure scientific information (i.e., science standing alone without context), withstanding the inevitable arrival and departure of personalities and stakeholders.

As we finish this story of collaboration and transformation in the Ozarks, it is evident that, although this topic is rife with scientific information and associated challenges/solutions, there is a perspective that can treat the entire situation of overcoming challenges on the Upper White River through reflection and celebration, focusing intently and predominantly upon the passions and deeds of the people in this place, with a relatively lesser focus on the science. We, as the authors of this chapter, felt it very important to intentionally emphasize the people of the Ozarks in this story, and not the science of the river system and associated landscape, not because one or the other is less important, but rather to recognize the potential for overemphasizing the science, and/or underemphasizing communities and their needs/perspectives. Unlike the standard narratives within best practice manuals of the past, we believe that, through the stories of people, we gain added dimensions to what is otherwise practiced in the same landscape, and indeed in the Ozarks, there is plentiful scientific information and associates best practices that have been developed and implemented. As one experiences the resolve of the people of the Ozarks, through their own experiences and words in this chapter, one realizes that science and best practices information in isolation could not have solved the problems of the Upper White River. For some special reason, a human reason, the passionate people of the Ozarks took hold and made a difference in the quality of life for the whole of their place, and the land itself.

As with the other stories along our journey of collaboration, in the Ozarks it happened… true transformation, and we celebrate this fact, simply by sharing the hundreds of "aha" moments, stories, and contributions of thousands of people who played their special roles, big or small, separate or together, in the healing process of a special place. Shaping and changing history, each of those who contributed to the healing process added their piece to the collective victory, for their water and for a history of the Ozarks that future generations can be proud of. Resourceful are they, the people, as always, to the end.

3.9 Memorializing a River through Its People

The history and involvement of the people of the Ozark Plateau necessitates special acknowledgements to the following individuals who helped the Upper White River forever: Loring Bullard, David Casaletto, Alan Fortenberry, Brian Haggard, John Moore, Holly Neill, Peter Herschend, Todd Parnell, Bob Pavlowsky, Tim Smith, and to the memory of Floyd Gilzow, John Lewis, Ben Parnell, and Duane Galloway, who inspired all of those who live on in their actions. Acknowledgements also go to those who elected not to have their names revealed across the watershed who offered up celebration and wisdom for the reader. A special acknowledgement to the hundreds of committed and knowledgeable delegates to the three watershed summits and hundreds of meetings held anywhere and everywhere in the Basin over the decades. It is also important to acknowledge all of the people in every village, city, county, utility, school, nonprofit, business, church, youth group, and state or federal agency who had their important moments, or seasons; sometimes in circumstances like those in the Basin, it may seem that you are the forgotten champions, but your conversations and contributions are anything but forgotten to those who have been to (or live in) this place, the Ozarks. And now those who can read your incredible story, along with those of us who have witnessed it, can appreciate and understand how all of your work mattered… and it always will.

Disclaimer

The U.S. Environmental Protection Agency through its Office of Research and Development funded and collaborated in the work described here. It has been subjected to the Agency's peer and administrative review and has been approved for publication. Mention of trade names or commercial products does not constitute endorsement or recommendation for use.

4

Integrating Traditional Knowledge and
Geospatial Science to Address Food
Security and Sustaining Biodiversity
in Yap Islands, Micronesia

Marjorie V. Cushing Falanruw

*USDA - Forest Service, Pacific Southwest Research Station,
Institute of Pacific Islands Forestry*

Reed M. Perkins

Queens University of Charlotte

Francis Ruegorong

Yap State Division of Agriculture and Forestry

CONTENTS

4.1 Introduction

Islands are ecosystems at a human scale and island societies are so integrated that scientists residing on islands are not just part of a community of scientist peers but part of the island community. This chapter provides an account of science mingled with the development of environmental efforts necessitated by islands' increased commerce with the wider world, and global challenges. It follows a sequence of scientific and social responses to general challenges of population growth, economics, climate change, and efforts to safeguard biodiversity, toward a more specific focus on integrating traditional knowledge and modern scientific technology to maintain food security and biodiversity services on the islands of mainland Yap.

Located in the tropical western Pacific, Yap State is a member of the Federated States of Micronesia (FSM), one of several nations in the Trust Territory of the Pacific Islands. The islands had been administered by Spain, Germany, and Japan in the past and are currently associated with the United States under a Compact of Free Association. Under this Compact, the United States is providing development assistance until 2023.

Yap State consists of a close cluster of high islands and fourteen atolls and small islands distributed over some 400,000 km² (154,441 mi²) of ocean (Figure 4.1). The outer islands collectively occupy about 26 km² (10 mi²), with the atolls having an average height below 2 m (21.5 ft) and with a maximum elevation of about 5 m (53.8 ft). "Mainland Yap" or simply "Yap," the main focus of this chapter, covers 100 km² (39 mi²) with the highest point being 174 m (571 ft). Yap islands lie within an area affected by the Asian monsoon, and the climate is wet with irregular dry seasons greatly affected by El Niño-Southern Oscillation (ENSO) phenomena. The climate is warm and humid with a mean annual temperature of 27°C (81°F) and an average annual rainfall of about 300 cm (118 in). The precontact population of mainland Yap has been estimated at five to six times as large as the 1980 census figure of 8,000 (Hunt et al. 1949; Underwood 1969). A period of depopulation occurred after foreigners settled on the island and continued through the end of the Japanese administration that ended with World War II (Useem 1946). Since 1946 the population has been increasing rapidly. The 2010 census reports the State's total population as 11,377 with approximately 65% living on Mainland Yap (FSM Division of Statistics 2012). By ethnicity, 52% of Yap's population are mainland Yapese and 41% are from the outer islands. Mainland Yap serves as the center for virtually all economic and governmental activities within the State, and is the location of Yap's banks, hospitals, 2-year colleges, and international airports (Falanruw 1994; Yap Statistics Office 2011; Perkins and Krause 2018).

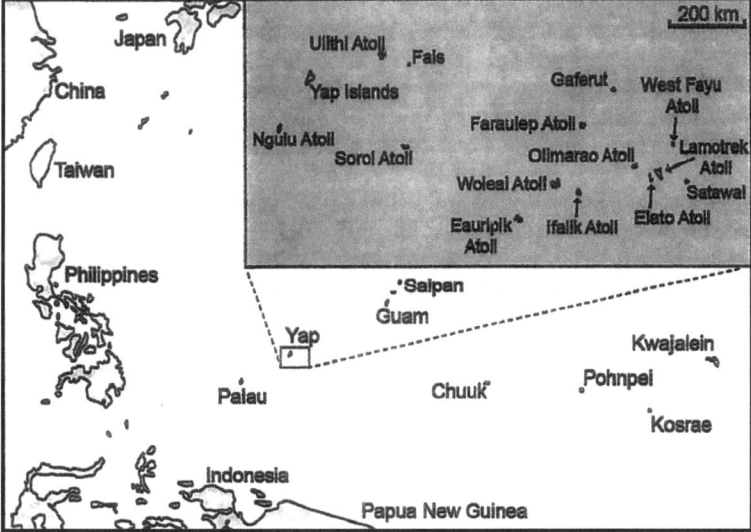

FIGURE 4.1

Map showing locations of Mainland Yap and outer islands of Yap State. Given predictions of sea-level rise, it is possible that all but the islands of mainland Yap and the outer islands of Fais will eventually become uninhabitable.

Yap is best known for its huge pieces of stone money that were mined in Palau and other areas and transported back by sailing canoes and rafts. These voyages would not have been possible however, had not the once dense population been supported by an effective food production system. In developing the system, the people of Yap modified natural communities in ways that imitated nature, but provided for people's needs. On Yap today, the stone money bespeaks a glorious past while aspects of the traditional food production system offer insights to a more sustainable future.

Yap's culture is tightly interrelated with the island's natural environment and social relationships were, and to a large extent still are, related to the allocation, sharing, and exchanging of natural resources (Falanruw 1996). The entire land and seascape was utilized in a food production system developed to feed the dense population of the past (Falanruw 1994). This system is superimposed on the natural topography and biotic communities of Yap and incorporates landscape architecture to manage the flow of water and nutrients in three general systems of agricultural production: *meliy* swidden gardens (hereafter *meliy*),[1] taro patch systems, and agroforests.

Meliy are generally initiated during the dryer season using controlled burning to open a small area, generally well under 0.02 ha (0.05 ac), in secondary vegetation or secondary forest. They may be intensively managed for prestigious *Dioscorea* yams or planted to yams and a variety of other crops. Harvests extend over a period of about 2 years after which the site is generally allowed to go fallow. *Meliy* gardening requires only a knife, a digging bar, and matches (or in the past a local flint stone and tinder), and provides a large harvest in proportion to a gardener's energy inputs as these gardens are subsidized by the ecosystem services of the forest fallow that has enriched soil fertility and structure. The making of *meliy* results in a mosaic of forest, secondary vegetation, and active gardens. This type of gardening is only sustainable if it does not exhaust forest and secondary vegetation resources. If the biomass of land has been exhausted, as in the case of Yap's savanna lands, more intensive systems of ditched garden beds must be utilized.

Water is managed in an extensive system of landscape architecture and "plumbing" throughout much of Yap, from ditched bed *meliy* in upper areas of watersheds, to areas where ditches are widened and used to grow taro in

---

[1] In this chapter, we will use the Yapese term *meliy* to describe gardens that in other publications have been termed "shifting cultivation," "swidden cultivation," "intermittent mixed gardens," or "slash and burn agriculture." Cairns (2007 and 2015) provides a comprehensive review of these important agricultural systems. Given the range of practices throughout the tropical world, we will utilize the term specific to Yap for these gardens *meliy*. The term "shifting cultivation" suggests that the gardening activities shift about. On Yap this is not always the case as a site that is within a family's resources may be used repeatedly but not continuously as it is generally necessary to allow the site to be fallow for a while. The term swidden agriculture is not often used in Micronesia, and the term "slash and burn agriculture" suggests careless burning of the forest. The Yapese use of fire to open garden areas is almost always controlled and limited in scope.

silt rich depressions serving as a variety of types of taro patches described in Section 4.6. Larger taro patches are developed in natural wetlands such as marshes, and in the past some areas of mangroves were converted to freshwater marshes and then into taro patches. Where more water collects in lands near the bottom of watersheds, low areas are deepened into taro patches and the excavated soil used to form raised garden beds, paths, and living areas. These raised areas are planted to a variety of useful trees resulting in Yap's tree garden taro patch agroforests. These agroforests are highly diverse in both species and varieties of food trees and other plants, and represent a considerable adaptation of a food production system to the landscape, and to microhabitats. Yap's agroforests serve as a supermarket, a hardware store, and a pharmacy for villagers while providing the ecosystem services of forests, and habitat for wild species as well.

4.2 Current Challenges

Today, Yap State and the FSM face three great challenges: declining financial aid, climate change and sea-level rise, and the need to sustain biodiversity and ecosystem services as the foundation of self-reliance. The details of these challenges are summarized in Sections 4.2.1–4.2.3.

4.2.1 Declining Financial Aid

FSM islanders face an uncertain economic future. The U.S. Trusteeship was replaced by a Compact of Free Association (COFA), originally signed in 1986, and then amended in 2003, to support vital program areas: education, health, infrastructure, public sector capacity building, and the environment. While the COFA economic assistance is beneficial in many ways, it has also created a relationship of economic dependency. The FSM currently receives approximately US$130 million per year (US Relations with the FSM 2017), or approximately 58% of total national revenues. Roughly two-thirds of employees nationwide work for the government (Central Intelligence Agency 2017), with limited options for other employment. As amended in 2003, direct financial assistance to the FSM, delivered through annual sector grants, will end in 2023. Prior to that date, an increasing percentage of U.S. aid funds are being placed in a trust, which after 2023 is intended to serve the same purpose as prior grant assistance (Gootnick 2016). The long-term self-sufficiency of this financial structure is doubtful. A 2016 Department of Interior financial review (U.S. Department of Interior 2017) indicates a "low probability" that the trust fund status after 2023 will be equivalent to a fiscal year 2023 grant aid (Perkins and Krause 2018). If Yap State follows the pattern of other developing countries, the most likely response to fill the financial gap will be an unsustainable exploitation of natural resources. This is already beginning to happen through activities such as the depletion of old growth tropical hardwood trees with introduced sawmills and foreign fishing ventures exploiting spawning aggregations for the live reef food fish trade (McClure 2018a, 2018b).

4.2.2 Climate Change and Sea-Level Rise

Given their small size, limited resources, fragile ecosystems, and suscepti-bility to natural disasters, Small Island Developing States (SIDS) are often characterized as particularly vulnerable to the impacts of climate change (United Nations 1994; Nurse et al. 2014). Satellite altimetry shows that sea levels in the area of the FSM have been rising by an average of 5–10 mm (.20–.78 in)/year since 1993—above the global mean of about 3 mm (0.12 in)/year for this period (Fletcher and Richmond 2010). Reported climate changes for the FSM include increasing air and sea surface temperatures, increasing numbers of tropical storms, and increasing ocean acidification (Australian Bureau of Meteorology and CSIRO 2014). Though precipitation across Yap is highly correlated with ENSO occurrence and intensity (Pacific Islands Climate Education Partnership 2014), The Australian Bureau of Meteorology and Commonwealth Scientific and Industrial Research Organisation (CSIRO 2014) report no discernible change in Yap's average monthly precipitation since 1952. Local gardeners however complain of changes in the day to day pattern of rainfall which results in conditions that are not conducive to traditional agricultural activities.

Saltwater intrusion is affecting coastal areas in both mainland Yap and the outer islands of Yap State. In Yap's outer islands especially, saltwater intrusion is impacting freshwater supplies and food production (Hezel 2009; Werner et al. 2017), and beach erosion is a major concern (Hezel 2009; Keener et al. 2012). The sensitivity of atolls, in general, to saltwater intrusion and drought is well documented (e.g., Terry and Falkland 2009; Chui and Terry 2015; Werner et al. 2017). Of particular concern is the coincident occurrence of extreme high tides, rising sea levels, and strong storm surge, potentially leading to the overtopping of atolls by saltwater (Werner et al. 2017). Within Yap State, saltwater intrusion during such an event in 2007 led to the mortal-ity of 90% of the taro crop on the islet of Falalop, Ulithi atoll, and 75% of the taro crop on the islet of Falalop, Woleai atoll (Hezel 2009).

With predictions of sea-level rise ranging from a conservative 0.2–0.6 m (0.6–2.0 ft) or more (International Panel on Climate Change IPCC 2007) to 0.6–2 m (2.0–6.6 ft) during the 21st century (U.S. National Research Council 2010), with greater and more abrupt rise possible (Meehl et al. 2007; Ananthaswamy 2012), the small size, coralline soils, and shallow freshwater lenses may result in atolls becoming uninhabitable in the next few decades (e.g., Nunn 2013; Storlazzi et al. 2015; Perkins and Krause 2018). The most likely place for outer island Yapese, who make up about 41% of the state's population, to relocate is to mainland Yap, but mainland Yap is also being impacted by climate change and intensive ENSO events. For example, during the strong El Nino event of 2015–2016, Yap experienced a record drought and wildfires that burned over 7% of the island.

Efforts to assess the vulnerability of Yap's food production and biodiversity to sea-level rise is hampered by a lack of more precise digital elevation models (DEMs) than can be derived from the 5 m (16.4 ft) contour intervals of the U.S. Geological Survey (USGS) map of 1983, which is based on aerial photographs taken in 1969 and field checked in 1980, with selected hydrographic data compiled from a DEM based on 1978 data. The very subtle changes and narrow range of island/atoll terrain changes, and the lack of precision obtainable from older maps compared to contemporary digital data points, make it difficult to reconcile locations of important features, obviating the critical need for much greater vertical and horizontal precision of elevation models. Such precision is currently obtainable from state-of-the-art Light Detection and Ranging (LiDAR) data, allowing for the generation of DEMs for both mainland Yap, which is especially needed for the outer islands of Yap that lie so close to sea level and are imminently in jeopardy.

4.2.3 Need to Sustain Biodiversity and Ecosystem Services

It is understood that biodiversity is important to maintaining stable ecosystems, especially in an era of climate change, and that food production is an important ecosystem service, among several. Neither Yap's soil nor waters are nutrient rich. The relatively high level of biodiversity seen in Yap is dependent on the recycling of nutrients and there is little surplus to export. The island's food production system is dependent on ecosystem processes and conservation of biodiversity is crucial to the preservation of Yapese culture, which depends largely on the islands' natural resources.

As the FSM was originally part of the United Nations Trust Territory of the Pacific Islands (TTPI), which focused on infrastructure, political development, and some economic development there was relatively little emphasis on sustaining biodiversity. Indeed, during the TTPI there was but one conservation officer for all of the TTPI, on an insular area that has now become the Republic of the Marshall Islands, the FSM, the Commonwealth of the Northern Marianas, and the Republic of Palau.

Much of the governmental structure of the TTPI was carried over into the newly organized nations that emerged, and as a result, there are few or no governmental positions focused on sustaining biodiversity and ecosystem services in the resulting entities. The world at large however has become concerned about biological diversity as a shared heritage and responsibility and at the Earth Summit in Rio de Janeiro, representatives of the international community came together and developed the Convention on Biodiversity (CBD), completed on World Environment Day, on June 5, 1992; the FSM became a signatory to the CBD on June 12, 1993.

Today the FSM and Yap State must somehow manage the impacts of economic challenges and climate change, while maintaining the health of the ecosystem that is the foundation of self-reliance.

4.3 Collaboration to Address Current Challenges

While it is often pointed out that many traditional practices of Pacific peoples such as resource apportionment, cultural restrictions, and customary tenure had conservation value, these practices are fading and modern technologies and markets are creating a need to consciously manage resources and protect the environment in new ways. It is necessary to not only conduct environmental science research but to also involve people and communities in projects, and to design studies so that they address people's needs.

The limited emphasis on sustaining Yap's terrestrial biodiversity inherited from the governmental structure of the TTPI has been augmented over the years through the efforts of nongovernmental organizations (NGOs), U.S. Forest Service programs, and collaboration with Queen's University of Charlotte. More recently the FSM's responsibilities as a signatory of the Convention on Biodiversity has led to local, national FSM, and international efforts assisted by The Nature Conservancy (TNC), Conservation International, and other groups with the local FSM Micronesia Conservation Trust dedicated to obtaining and managing conservation funds for all of Micronesia.

4.3.1 Early Efforts

The world's second Earth Day dawned with Yap High School's observance of the Day in 1971, and Yap's first environmental NGO, the Yap Institute of Natural Science (YINS) was chartered in 1975 under the TTPI. It was dedicated to the collection of ethnobiological knowledge and research in natural history, adaptive technology, eco-development, and the ideal of maintaining indigenous integrity through the sustainable use of local resources. With very limited means, the small NGO, YINS, undertook modest local projects and also collaborated with the Smithsonian Institution and the South Pacific Commission (now Secretariat of the Pacific Community), and the International Union for Conservation of Nature (IUCN) on a number of larger projects (e.g., Fosberg and Falanruw 1974; Fosberg et al. 1975; Falanruw 1979), and with Yap Government and communities (such as Falanruw 1996). From 1980 to 2014 YINS produced a Yap Almanac Calendar (Falanruw 1980–2014), with tides, moon phases, and other astronomical events, holidays, seasonal phenomena, information on biodiversity and natural history, ethnobiology, adaptive technologies, and eco-developments to publicize and celebrate a growing number of environmental and conservation efforts.

4.3.2 Yap State Forestry and U.S. Forest Service Programs

Yap State Forestry within the Division of Agriculture and Forestry (Yap Forestry) works directly with the Public on State and Community projects and oversees a local annual cycle of eco-events in partnership with other government, NGO, and community groups, flying the Earth Flag on Earth Day for most of the past 47 years. The work of Yap Forestry is augmented by the U.S. Forest Service Pacific Southwest Region, Institute of Pacific Islands Forestry (IPIF) research, and State and Private Forestry (S&PF) programs.

Research published by the IPIF includes a series of vegetation maps of the high islands of Micronesia (MacLean et al. 1986; Whitesell et al. 1986; Cole et al. 1987; Falanruw et al. 1987a, 1987b, Falanruw 1989a); research on fruit bats (Morse et al. 1987; Falanruw 1988; Falanruw 1989b); mangrove and carbon sequestration by Yap's vegetation (Donato et al. 2011, 2012); soils (Falanruw et al. 1987b); a dictionary of local, common, and scientific names of trees and shrubs of the Mariana, Caroline, and Marshall Islands (Falanruw et al. 1990); culture and resource management (Falanruw 1991); and a field guide to the trees of Yap (Falanruw 2015). Other publications of the IPIF in collaboration with YINS include floristics (Falanruw 1989), birds (Pratt et al. 2008), reptiles (Wynn et al. 2012), fruit bats (Falanruw 1989; Falanruw and Manmaw 1990; Wiles et al. 1991), as well as traditional and modern environmental management (Falanruw 1985, 1990a, 1990b, 1992, 1997, 1999; Keldermans et al. 1994).

IPIF research is complimented by the S&PF programs that provide grant assistance to Yap Forestry. The first S&PF program to be initiated was the Urban and Community Forestry (U&CF) Program. The first Yap State U&CF plan (1977) stated, "Previous generations of Yapese developed the surrounding landscape into a food production and living system and that landscape in turn sustained Yapese culture. It is important to maintain and enhance this connection while progressing into the future. There is need to foster an improved environment for all people in Yap by organizing and encouraging the planting and maintenance of trees so that communities will be cooler, have cleaner air and water, quieter streets and paths, more peaceful neighborhoods, improved nutrition, sources of materials for artisans as well as building materials and medicines; stronger village economies and more pleasant surroundings in the places where we live, work and play." The program developed and successfully implemented three successive 5-year plans and now operates under the State-Wide Assessment and Resource Strategy (FSM 2010). The U&CF and other S&PF programs are integrated with FSM national, state, and community efforts.

The Forest Stewardship Program encourages the long-term stewardship of nonfederal, nonindustrial private forest lands. Goals of the program include strategies in the Forest Action Plan (FAP), especially in important forest resource areas; outreach through a landscape stewardship approach; improved water quality and supply; mitigation and/or adaptation to climate change; and support for community planning and capacity building for these efforts. This program helps to support the Yap Forestry nursery that can accommodate more than 6,000 tree seedlings at any given time. It has also supported demonstration projects and technical assistance to private land owners for projects such as savanna reforestation, timber, native and food tree planting, and the development of forest stewardship plans. A Forest Health Program also assists in addressing forest pest and diseases as well as combating invasive species.

The U.S. Forest Service's Wildfire and Aviation Cooperative Fire Protection program (Cooperative Fire) assists with wildfire prevention and suppression training, providing surplus fire vehicles for use in fire suppression by the Yap Fire station and works with communities to develop Community Wildfire Protection Plans (CWPP). Three successive Yap State Public Safety Fire Section Wildfire 5-year plans have been successfully completed along with CWPP plans. The program currently supports Yap Forestry work with communities to develop shaded fuel breaks to decrease contiguous areas of flammable vegetation in savannas in order to reduce the extent of wildfires. Upon completion of their CWPP, Tomil Municipality (Mainland Yap) underwent training in wildfire suppression using hand tools supplied by the program. The CWPP and training resulted in cooperative efforts of the government fire department and community for the first time during the extreme wildfire season of 2015–2016. Figure 4.2 shows the Tomil community trainees with an insert of one of the trainees using a backpack sprayer to suppress flames. The efforts protected

FIGURE 4.2
Tomil community members trained in wildfire suppression in front of traditional municipal *pebay* building. Inset: Community firefighter suppressing fire with backpack sprayer provided by the U.S. Forest Service's Wildfire and Aviation Cooperative Fire Protection program.

community structures, yet there were not enough resources to suppress an extensive wildfire associated with the record drought and the presence of large contiguous areas of fire prone savanna vegetation. Now, with assistance from Cooperative Fire, Tomil and other communities are developing shaded fire breaks to break up large contiguous areas of flammable vegetation.

4.3.3 The Environmental Stewardship Consortium

At the First Micronesian Traditional Leadership Conference held in Palau in 1999 traditional leaders of Micronesia declared, "We are mindful that our environment and our natural resources are all important, for they are the foundation of our economies, our cultures, and our identities as Pacific Islanders." Upon return to Yap the Council of Pilung (Chiefs) mandated the development of an environmental group to work cooperatively with Yap State Government to develop an environmental stewardship program for Yap State. Subsequently, an Environmental Stewardship Consortium (ESC) was pulled together by the late director of the Yap Community Action Program (Yap CAP).

As a signatory to the Convention on Biological Diversity, the FSM needed to develop a Biodiversity Strategy and Action Plan (BSAP), and FSM States were asked to develop draft BSAPs to contribute to the National plan to protect and sustainably use life on earth, and in the FSM. The task of compiling the Yap State BSAP was taken on by the newly organized ESC.

4.3.4 The Queens University/Yap State GIS Program

In 2000, Dr. Reed Perkins of the Queens University of Charlotte, North Carolina (QU), and co-author of this chapter, approached the ESC about a cooperative program. Queens University guarantees its students an overseas experience as part of their education. Most students visit Europe and other established areas. Dr. Perkins had previously worked in Micronesia and felt that it would be good if students could not only visit Micronesia but help address local needs. The ESC was in need of assistance with GIS, and 17 years of cooperative work between QU and Yap State (QU/Yap) followed. Groups of about a dozen students and one or two professors began coming to Yap, living in local accommodations, and conducting geospatial data-oriented fieldwork in partnership with Yap CAP and government agencies. As part of these efforts, QU brought needed equipment and set up a global positioning system (GPS) base station in order to obtain more accurate GPS readings. One of the early findings of the project is that the features based on USGS mapping (USGS 1983), while internally correct, differed about a half mile (eight-tenths of a kilometer) from actual locations, as evidenced by updated GPS measurements. This led to appropriate adjustments to the project's GIS database that was eventually housed at the Yap State GIS lab at the Division of Land Management. Students were carefully selected for the program and underwent special preparation prior to travelling to Yap. The first group to come to Yap digitized a vegetation map (see Section 4.5.1) that was utilized to support IPIF research, an experiential element of the educational program for students from Dr. Perkin's lab. This work greatly enhanced the usefulness of the map, and enabled further GIS analyses.

The QU/Yap State program carried out many projects over the years, including the mapping of areas burnt by wildfires, areas infested with invasive species, areas designated as having special biodiversity or cultural significance, and the integration of taro patches and *meliy* gardens within forested areas of Yap. The mapping of areas burnt by wildfires contributed to Yap's annual reports to the Cooperative Fire program. The maps also revealed wildfire patterns such as repeated fires in areas of savanna and the pattern of chronic savanna wildfires of limited area on most years with moderate dry seasons and more acute and widespread wildfires that also damage forest resources on years with more severe droughts. Fire maps were initially made by walking the perimeters of burnt areas with GPS instruments, and are currently made using GoogleEarth satellite imagery in concert with local records of wildfires. Experience and data from the project have been used in Cooperative Fire sponsored projects in wildfire suppression, the development of CWPPs, and ongoing projects on shaded fuel breaks.

The QU/Yap State program also assisted the state's successful program to control a large infestation of *Imperata cylindrica*, the infamous *lalang* or *cogon* grass that is one of the world's worst invasive species. This USFS S&PF supported program was sharpened through the use of the GIS maps to chart progress, and documented the importance of continuity of control efforts. The GIS maps from this project showed how the initial infestation of 24.3 ha (60 acres) was reduced to less than a 0.5 ha (1.2 acres) by early 2004. The control program was then interrupted as a result of typhoon Sudal in April 2004, which destroyed Yap State Forestry vehicles and diverted staff to work on other more immediately urgent relief efforts. Unfortunately, the typhoon also damaged the trees planted to shade out regeneration of the *Imperata* grass. By early 2006 when control efforts were resumed, the grass had rebound to 6 ha (15 acres). By 2007 it was reduced to less than 1 ha (2.5 acres) and is now virtually eliminated.

Data collected by the QU/Yap State (GIS) program became the initial database of the Yap State GIS system, established within the Division of Land Management. Maps made with the use of the database continue to contribute to a wide range of applications including efforts to define areas of special biodiversity significance, as well as the watershed project described in detail in this chapter.

4.3.5 The FSM National and Yap State Biodiversity Strategy and Action Plans

The Yap State BSAP (Gaan et al. 2004) was first drafted in 2001 for input into the national FSM BSAP, and then finalized in 2004 to include recommendations and priorities of the National FSM Biodiversity Strategy and

Action Plan (NBSAP). The Yap Biodiversity Strategy and Action Plan (YBSAP) was developed by the Yap State Environmental Stewardship Consortium (ESC), organized by the late Director of Yap CAP, working in collaboration with representatives from the Government agencies of Forestry of the Division of Agriculture and Forestry; Marine Resources Management Division, Commerce and Industry, Offices of Planning, Education, and Tourism; the Yap State Environment Protection Agency; and College of Micronesia. NGOs included the Yap Community Action Program, the Yap Institute of Natural Science and Yap Women's Association. Communities were represented through municipal representatives designated by the Council of Pilung and Council of Tamil, the Chiefs of Mainland Yap, and the neighboring outer islands respectively. Meetings of the ESC were open to all interested parties.

The FSM Biodiversity Strategy and Action Plan (FSM NBSAP 2002) points out that the FSM's biodiversity is the nation's living wealth and is important to every facet of Micronesian life and is the basis for economic development. It also recognizes that given changes in lifestyles, commercial activities, pollution, and the global environmental crises of climate change and sea-level rise, the biodiversity of the FSM is under increasing threat. The document also makes it the prerogative of each State of the FSM to enact their own legislation and address all issues relating to conservation of biodiversity. Thus, the implementation and monitoring programs of the NBSAP are to be undertaken by individual states of the FSM (Mace 1999; FSM NBSAP 2002).

The YBSAP, like the NBSAP, described the State's biodiversity, including the areas of special biodiversity significance (ABS) recognized in the Blueprint for Conserving the Biodiversity of the FSM, threats to this biodiversity, and actions needed to utilize these natural resources on a sustainable basis. In recognition that there were only about 14 Yap State and semigovernmental staff available to address the 198 actions of the NBSAP in addition to their ongoing work, a strategy of turning conservation from the work of a few small agencies and NGOs into a community concern was adopted. Accordingly, local friendly metaphors such as Figure 4.3 were utilized to illustrate principles. The overall effort of "Taking Care of Yap" was translated as *Chothowliy yuu Waab* and presented using a local metaphor of a dewdrop in a taro leaf. In the past, these precious drops of pure water were collected for use in medicines and for babies, and it was pointed out that they, like precious babies, must be handled with care lest they slip away and be lost. Yap was likened to this dew drop in a taro leaf and this became the logo of Yap's environmental efforts.

In 2005, the work of the Yap ESC resulting in the YBSAP, results from the QU/Yap State GIS program, and other activities were reported at simultaneous meetings of the Environmental Protection Agency Pacific Islands Environmental Conference (Falanruw and Chieng 2005) and the third Micronesian Traditional Leadership Conference (Falanruw 2005). At the same time, the metaphor of "The Third Nguchol" (Figure 4.3) resonated in other Micronesian cultures and was presented at the Environment conference in Pohnpei by the late Director of Chuuk State Environmental Protection Agency.

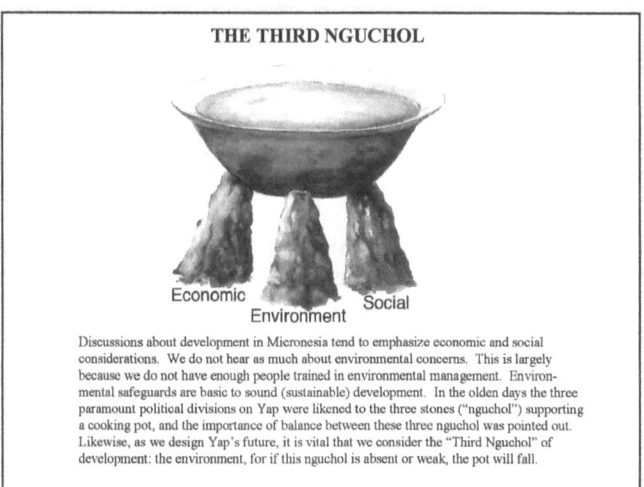

THE THIRD NGUCHOL

Economic   Social
Environment

Discussions about development in Micronesia tend to emphasize economic and social considerations. We do not hear as much about environmental concerns. This is largely because we do not have enough people trained in environmental management. Environmental safeguards are basic to sound (sustainable) development. In the olden days the three paramount political divisions on Yap were likened to the three stones ("nguchol") supporting a cooking pot, and the importance of balance between these three nguchol was pointed out. Likewise, as we design Yap's future, it is vital that we consider the "Third Nguchol" of development: the environment, for if this nguchol is absent or weak, the pot will fall.

FIGURE 4.3
The Third Nguchol, artwork by Luke Holoi from Falanruw (1996). A goal of the FSM biodiversity strategy and action plan is to protect and sustainably manage a full representation of the FSM's marine, freshwater, and terrestrial ecosystems.

4.3.5.1 *The Blueprint*

Augmenting the development of the FSM NBSAP (FSM 2002), the FSM States and National Government, with assistance from TNC, USFS, the U.S. Department of the Interior, and the UNDP Global Environment Fund developed a Blueprint for Conserving the Biodiversity of the FSM using TNC's methodology. The project was carried out over a period of 2 years and involved consultations with Government agencies, scientists, local experts, NGOs, and communities. The "Blueprint" identified areas of special biodiversity significance termed "ABS" (Areas of Biodiversity Significance). The USFS IPIF vegetation maps of high islands of Micronesia (MacLean et al. 1986; Whitesell et al. 1986; Falanruw et al. 1987a, 1987b, 1989; Cole et al. 1987) were used to help define terrestrial ABS.

4.3.5.2 *The Micronesia Challenge*

The Blueprint was followed by the Micronesia Challenge (MC) (The Nature Conservancy 2006). The MC was initiated by TNC and the President of Palau, and on March 1st, 2006 the FSM signed the Declaration of Commitment to the Micronesia Challenge. The declaration commits the signatories to, "effectively conserve at least 30% of the near-shore marine and 20% of the forest resources across Micronesia by 2020." Signatories to the 2006 Declaration included the Republic of Palau, FSM, Republic of the Marshall Islands, the Commonwealth of the Northern Marianas, and Guam (Chief Executives of

Micronesia 2006). The MC covers nearly 5% of the Pacific Ocean and 7% of its coastlines and will collectively protect a marine area equal to the size of the Gulf of Mexico. The MC was launched at the Eighth Conference of the Parties to the Convention on Biological Diversity held in Brazil in March 2006. Recognizing the unique contribution of islands to global biodiversity, over 180 countries adopted the Island Biodiversity Programme of Work, which lays out guidance for island nations and nations with islands for integrated conservation and management of their vital resources. The TNC and Conservation International each pledged US$3,000,000 dollars toward conservation across Micronesia. The Global Environment Facility has pledged a US$6,000,000 match and island leaders and partners are working to secure additional funding. It is estimated that an endowment of approximately US$100,000,000 will need to be raised to support the long-term sustainability of the MC in all five nations (TNC 2006). In response to the Micronesia Challenge, financial support for marine conservation efforts became available especially as a result of the International Coral Reef Initiative. Initial support for land stewardship was largely from the ongoing U.S. Forest Service S&P program.

4.3.5.3 The FSM State-Wide Assessment and Resource Strategy

In 2010, the FSM National Department of Resources and Development, with guidance and technical assistance from the U.S. Forest Service, Pacific Southwest Research Station's Institute of Pacific Islands Forestry (IPIF) developed the FSM State-Wide (Forest) Assessment and Resource Strategy (SWARS) (FSM 2010). The SWARS is "a tool for islands to identify their highest priorities for forest resource management and seek implementation of their strategies with on-island partners with assistance from the US Forest Service." It provides a basis for subsequent annual grant proposals, including opportunities to apply for competitive grants to augment the ongoing grant program.

The forest assessment section of the Yap chapter of the SWARS made use of GIS (i.e., maps and other geospatial data analyses) and the analyses made possible by the QU/Yap State program. The document also drew on other plans such the Yap State Division of Agriculture and Forestry 5-Year Plan (2009), the Yap State 5-Year Wildfire Plan (2009), and the Yap State Invasive Species 5-Year Plan (2009). Each of these plans was developed in consultation with communities, NGOs, and relevant government agencies.

As part of the SWARS, each FSM State was asked to prioritize its forestry issues. The top three priority areas defined in the Yap State chapter of the SWARS (now called the Forest Action Plan [FAP]) were Food Security, Biodiversity Conservation, and Watersheds. The report describes long-term desired conditions, threats, available resources, and strategies. Given the state's limited forestry staff and resources, the general strategy for achieving desired conditions recognized a need to "increase Public and Community awareness and the means to protect the ecological integrity of Yap while providing for food supplies and healthy livelihoods, and to turn Forestry, from a small government agency, into a community concern" (FSM 2010).

4.4 Bringing Things Together for Action in Place: The Yap State Watershed Project

In 2012, Yap Forestry submitted a successful proposal for a competitive grant to augment the ongoing S&P grant. The proposed project utilized a watershed approach to food security and sustaining biodiversity that included the use of GIS technology. The project was implemented in 2013–2015. In keeping with the Yap State SWARS priorities of food security, biodiversity conservation, watersheds, and development of local capacity, the goal of the project was "to enable individuals and communities to understand their place in relation to watershed processes and to provide tools to enhance food security while protecting biodiversity in the face of climate change." The results of the project are summarized below and discussed in more detail in later sections.

4.4.1 Overview of Watershed Project

The project took a three-pronged approach to evaluate Yap's watersheds and their stewardship: (1) GIS mapping and analysis, (2) field interviews, and (3) village projects. GIS technology was used to analyze attributes of Yap's watersheds in parallel with a review of relevant literature, field interviews, and projects on traditional Yapese methods of using watersheds. One hundred and eighty-nine watersheds and interfluve areas were delineated and identified so that data on attributes relevant to food security and biodiversity could be mapped and gathered into a database for each watershed to enable communities to evaluate their watersheds and prioritize watershed stewardship activities.

Field interviews on traditional practices in watersheds were carried out throughout Yap to gather information on traditional use of watersheds and the presence of taro patches, and shifting (intermittent) gardens. Reports on the location and occurrence of salinization of 2,707 taro patches, and locations of 1,959 sites of recent intermittent gardens were digitized in order to evaluate food production capacity and its vulnerability to salinization and increased incidence of storm surges of ocean water, as both are understood to be associated with sea-level rise. Overlaying the assessment information maps with a digital elevation model (DEM) indicated that the impacts of salinization were reported for elevations up to 5 m (16.4 ft), much higher than expected. A general GIS analysis based on slope and soil types indicated that 68% of areas suitable for taro production are located at or below 5 m (16.4 ft) elevation.

Community watershed projects undertaken as part of the project demonstrated the efficacy of managing stream flow to renovate taro patches. The projects also demonstrated a need to augment traditional methods with the use of modern materials in order to accommodate increased water flow from especially heavy rainfall (experienced during the project). This consideration is important because climate change predictions indicate continued increases in periods of heavy rainfall for the foreseeable future. Field interviews and observations revealed a number of traditional technologies for managing water flow and moisture in taro patches and shifting gardens that bear further documentation and development to enhance food security. Field observations also indicated negative impacts on taro patches and watersheds resulting from road projects.

Areas designated as having special biodiversity value (ABS) were demarcated so that watershed stewardship plans can include their protection and development into terrestrial protected areas in accordance with Yap's criteria for terrestrial protected areas under the Micronesia Challenge (Yap State 2014). Results of the project were incorporated into a watershed framework that includes a series of resource maps and a geodatabase that communities can use to evaluate their watersheds, prioritize watershed activities, and develop watershed stewardship plans and data-based proposals for projects. The following sections discuss detailed project results regarding food security.

4.4.2 Watersheds and Food Security: Agroforests

Almost all of Yap's food production is from the "nature-integrated" traditional agricultural systems briefly described earlier. Yap's tree-garden/taro-patch agroforests are a major sustainable source of food that also provide the ecosystem services of forests and habitat for many wild species, such as the endemic Yap monarch, *Monarcha godeffroyi* and other birds and endangered endemic fruit bats. Over 30 species of food trees are commonly found in Yap's agroforest with numerous varieties of the more important species such as coconut *Cocos nucifera*, breadfruit *Artocarpus* spp., and *bu'oy* nuts *Inocarpus fagifer*. There are especially many varieties of bananas such as *Musa* spp., taros *Colocasia esculenta*, *Cyrtosperma meruksii*, and yams *Dioscorea* spp. This great agrobiodiversity that is being lost even before it has been fully documented is very important to the resilience and adaptation of Yap and other islands to a changing climate.

Agroforests were recognized and demarcated in the vegetation maps of high islands of Micronesia, produced by the IPIF between 1986 and 1989 (MacLean et al. 1986; Whitesell et al. 1986; Cole et al. 1987; Falanruw et al. 1987a, 1987b, 1989); maps from this study revealed that agroforests are a major type found throughout the high islands of the Carolines. The 1987 vegetation map of Yap indicated that some 26% of the island was covered with agroforest. A number of subtypes of agroforest or of vegetation including an agroforest component were delineated in order to better understand the nature of Yap's agroforests and their dynamics. Subtypes of agroforest that were delineated included the following:

- General agroforest
- Agroforest with swamp forest inclusions
- Agroforest with secondary vegetation inclusions
- Agroforest in which the secondary vegetation included a significant bamboo component
- Agroforests that included over 20% coconuts
- Coconut groves that were developing into agroforest
- Urban and upland forests with agroforest inclusions

The digitation and color coding of the vegetation map as part of the QU/Yap GIS project enables both visual analog and digital GIS analysis of Yap's agroforestry resources. The development of a digital elevation model (DEM) makes it possible to approximate where agroforests are vulnerable to storm surge and sea-level rise, and the delineation of watersheds enables communities to evaluate their agroforest resources. Inasmuch as agroforests provide food resources as well as other ecosystem services, they were made eligible for inclusion as terrestrial protected areas under the MC as discussed in a later section.

4.4.3 Integrating Traditional Knowledge and GIS Technology: Taro Patches and *Meliy*

Assessing the extent of Yap's other systems of food production is more challenging as locations of taro patches and *meliy* gardens are generally obscured by tree canopies and other vegetation on aerial and satellite imagery. Locations of taro patches and *meliy* gardens were therefore determined by field surveys conducted by Ms Bernie Mininug of Yap State Department of Agriculture and Forestry, August 2013 to November 2014. A total of 176 discussions occurred with 115 village residents familiar with 117 of 120 of Yaps currently inhabited villages. Community members were chosen based on their knowledge about taro patches and *meliy*. Using section maps (approximate scale 1:5,000) of the 1983 USGS quad map (e.g., Rumung, Maap, Gagil-Tamil, Fanif-Weloy, and Dalipebinaw-Gilman quads) and corresponding printouts of satellite imagery of these areas, Bernie and village residents marked approximate locations and types of taro patches using roads, streams, and coastlines as geographic references. In addition, those taro patches that had been affected by salt water were marked.

A digital photo was then carefully taken of each paper section map. These photos were geo-referenced in ArcGIS 10.2 using road intersections and unambiguous coastal points as reference points. Heads-up digitizing was then used to enter the taro patches and saltwater intrusion points into the geodatabase. To increase the accuracy of placement (i.e., relative to streams, roads, and visible landscape features), the geo-referenced maps were displayed at 50% transparency over WorldView (©DigitalGlobe) imagery. Smaller taro patches were entered as point features and large taro patches were entered as polygons. Those taro patches, or sections of large taro patches that had been affected by salt water, were colored red. The attribute tables for the data on taro patches also indicates the type of taro patch. Validation assessments estimate that the resulting GIS map captures about 95% of large *muquot ni ga* taro patches, and about 80% of three other types of significant taro patches.

Yellow triangles represent an area where *meliy* gardens have been made in recent times and generally represent a cluster of gardens rather than individual gardens. In some cases, such as the hills behind Waanyan and Gachepar villages, there were more sites than could be reasonably recorded. It is estimated that the GIS map captures about 70% of the areas where *meliy* have recently been made. The mapped presence of ditched beds visible in savanna areas (Falanruw et al. 1987) and observed presence of ditched beds below a forest canopy indicates that in the past many more areas were gardened. Figure 4.4 provides a view of the distribution of taro patches and *meliy* gardens on Yap. Figure 4.5 is an enlargement of the area of Gagil-Tomil.

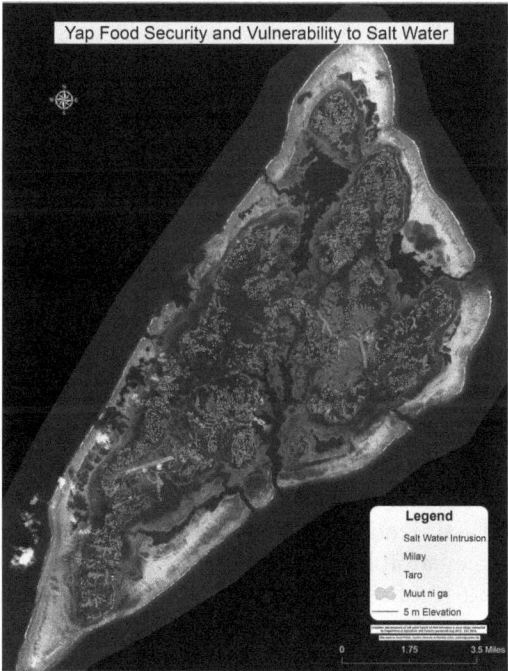

FIGURE 4.4

(**See color insert**) Map of taro patches and *meliy* gardens throughout Yap. Circles represent taro patches; yellow triangles represent *meliy* gardening areas; red circles and portions of polygons represent taro patches affected by salt water.

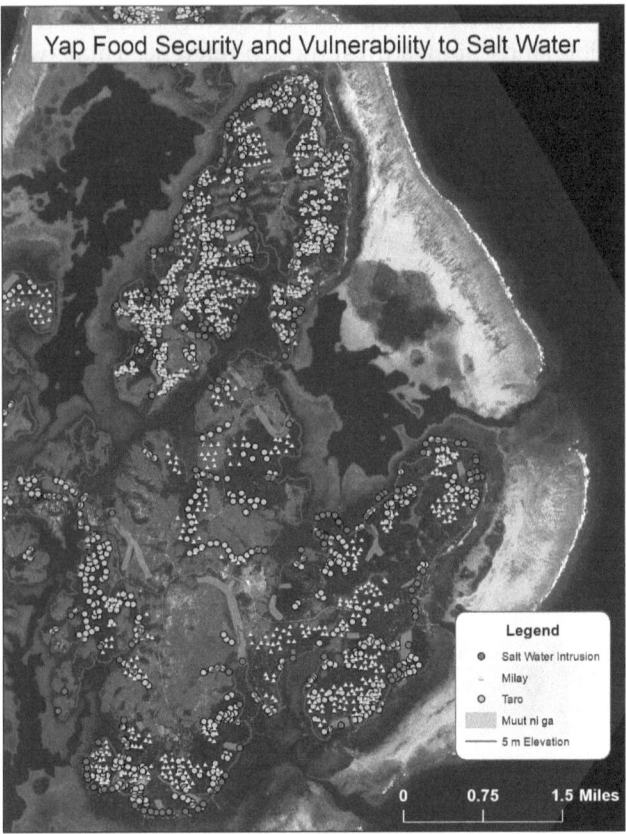

FIGURE 4.5
(See color insert) Enlarged view of Gagil-Tomil and map section of Yap. Round circles and polygons represent taro patches; yellow triangles represent *meliy* gardens; red circles and portions of polygons represent taro patches affected by salt water.

By overlaying the GIS map of taro patches and *meliy* with the digital elevation model (DEM), it was determined that a majority of taro patches and lower portions of a number of large taro patches were at or below a 5 m (16.4 ft) contour level. The GIS analysis also revealed that while highly spatially variable, and salinization was reported to have affected taro patches up to the 5 m (16.4 ft) elevation level. In addition, there were many taro patches lying within 2 m (6.6 ft) of sea level that were not reported to be affected by salt water. Section 4.5 discusses some traditional practices protecting taro patches from saltwater intrusion.

Surveys conducted after Yap's most recent severe storm, typhoon Sudal, reported a storm surge of 4 m (13.1 ft) in southern Yap. Eyewitness accounts indicate higher surge plus waves in specific areas. A GIS analysis of general areas deemed most suitable for taro production, based on slope and soil type, indicated that some 68% of the area suitable for taro production lies within 5 m (16.4 ft) of sea level. This analysis, combined with the reports that many taro patches occurring in this zone are already affected by saltwater, raises concerns for Yap's food security, especially at a time when residents of Yap's low-lying outer islands, who make up about 41% of the state's population, are migrating to the higher lands of mainland Yap, which puts further pressure on the food supply.

4.5 Overview of Traditional Management of Watersheds for Food Production and Modern Impacts

The delineation of both general and subtypes of vegetation in the 1987 map of Yap's vegetation combined with field observations prior to and as part of the watershed project makes it possible to model the general pattern of Yap's food production system and its dynamics. These models are presented in Section 4.5.1 followed by a more detailed account of traditional management in different sections of Yap's watersheds, as well as observations of modern impacts on these areas.

4.5.1 Overview of Yap's Vegetation

The vegetation of Yap has been much modified as a result of the agricultural system employed to feed the large population of the past. Figure 4.6 provides a color-coded view of this vegetation, developed through photo-interpretation of black and white aerial photographs from 1976 at a scale of 1:10,000 (Falanruw et al. 1987), digitized and then color coded as part of the QU/Yap GIS program. In the map produced, 26% of the island consists of tree garden/taro patch agroforest; about 34% of the vegetation on island consists of a mosaic of native forest, secondary forest, secondary vegetation, and active garden plots; and another 22% of the interior island is savanna land, with (coastal forested wetland) mangroves making up about 12% of the island's area. Other types of land cover in aggregate make up less than 6% of the remaining land area. Given the size of the map depicted in this book's format, many of the vegetation subtypes are not distinguishable; however, major vegetation types in the map are sufficiently distinct for the purposes of this discussion.

4.5.2 Traditional Land Use Zones

Yap's watersheds were, and to a large extent still are, traditionally managed from ridge to reef, as shown in Figure 4.7, which depicts a color code that corresponds with the vegetation map in Figure 4.6. The vegetation dynamics

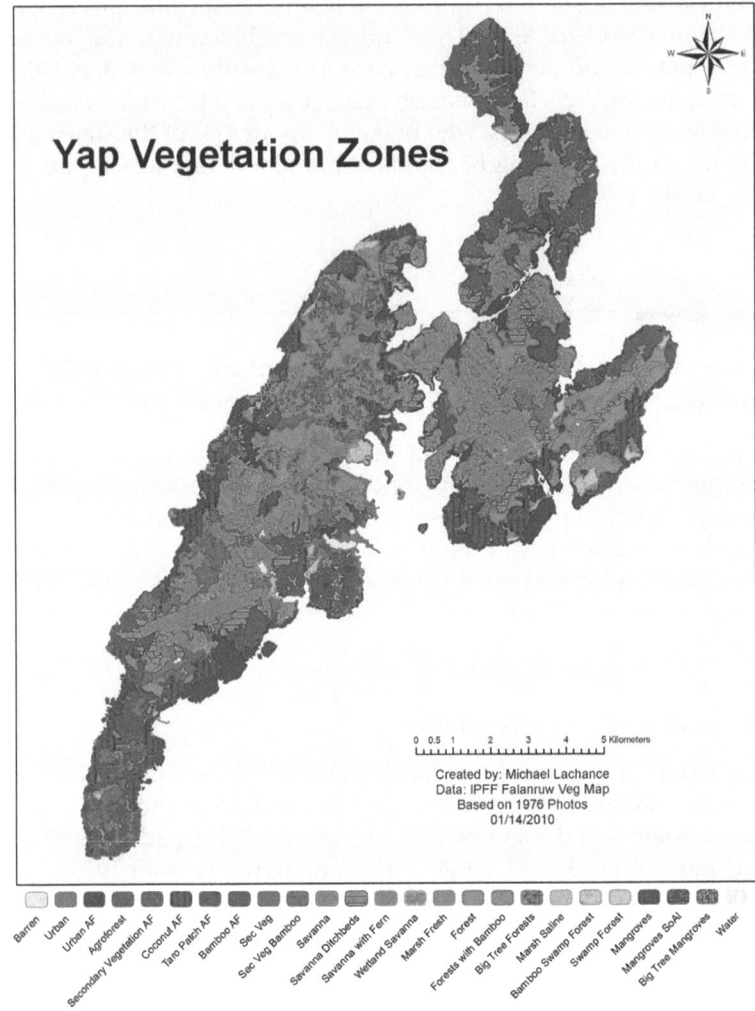

FIGURE 4.6
(See color insert) The vegetation of Yap based on data from Falanruw et al. (1987), color coded
to show major types and some important subtypes necessary for decision-making.

of the system are shown in Figure 4.8. Dialog with village residents served to
verify this diagram of land use; note that the arrangement of zones may vary
in different villages, depending on topography.

In Figure 4.7, native upland forest occurs at the top of the watershed and
in valleys. Savanna lands, often with ditched bed gardening areas, occur
at the top of the watershed or on slopes. Active *meliy* and secondary forest
and vegetation generally occur on lower slopes, while village tree-garden/

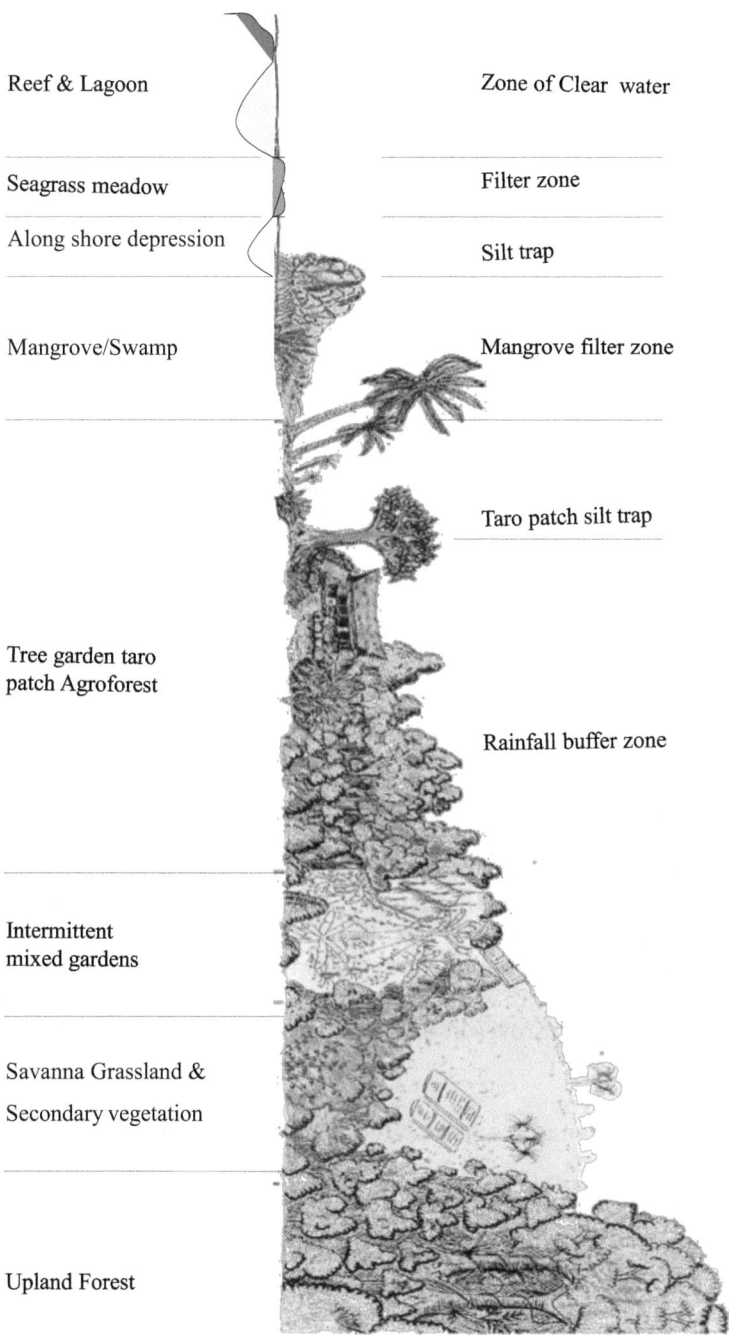

Reef & Lagoon

Seagrass meadow

Along shore depression

Mangrove/Swamp

Tree garden taro
patch Agroforest

Intermittent
mixed gardens

Savanna Grassland &
Secondary vegetation

Upland Forest

Zone of Clear water

Filter zone

Silt trap

Mangrove filter zone

Taro patch silt trap

Rainfall buffer zone

FIGURE 4.7
(**See color insert**) Diagram of traditional land-use zones. (Adapted from Falanruw (1995).)

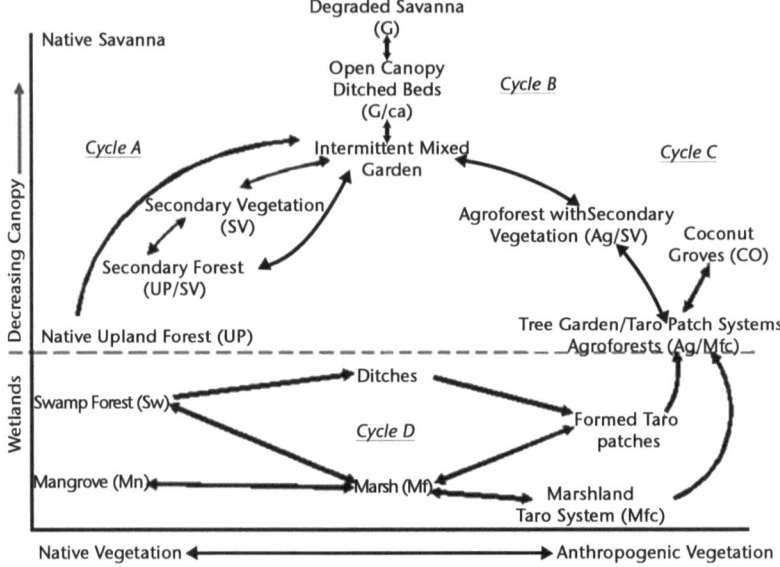

FIGURE 4.8
Hypothetical sequences of long-term vegetation dynamics involving traditional agricultural systems on Yap. The vegetation type codes (such as UP for upland forest) correspond to the type codes used for vegetation types demarcated in the work of Falanruw et al. (1987). (Diagram adapted from Falanruw (1995) and Falanruw and Ruegorong (2015).)

taro-patch agroforest surrounds villages. The system manages the flow of water so that silts are dropped into taro patches, acting as silt traps, and subsequently serving as a rich growing media, and/or the silt passes into mangrove areas. Mangroves filter runoff, serve as a detrital ecosystem that contributes to fisheries, and also protect coasts. In many areas, formerly managed coastal depressions beyond mangroves called *lupuw* serve as a final silt trap that transports silt out of the lagoon, where seagrass beds serve as a filter between mangroves and clear waters of the lagoon, reefs, and the open sea.

4.5.3 Vegetation Dynamics Associated with Food Production

Palynological studies indicate that Yap was mostly forested prior to human occupation (Dodson and Intoh 1999). The diagram below charts the vegetation dynamics likely to have occurred as Yap's population increased: from native vegetation on the left to anthropocentric vegetation developed to meet people's needs on the right. The initial molding of the landscape occurred during precontact times when the population of Yap was very high. Time lapse glimpses of Yap's vegetation are provided in the works of Tetens (1958), de Oca (1893), Volkens (1901), Johnson et al. (1960), and Falanruw et al. (1987), and were used along with field observations to develop Figure 4.8. Some of the changes such as the conversion of native forest and secondary forest to *meliy*, the conversion of mangroves to marshes and reversion of *meliy* to secondary vegetation, and the conversion of agroforests to agroforests with secondary vegetation have been observed over the past 50 years. In addition, one of the more significant changes that have occurred has been the general "leveling" of Yap as the landscape architecture of ditched and raised garden beds and managed taro patches has been less intensively tended to by contemporary generations.

Much of the native upland forest has been used for intermittent gardens, and very little of the mature native forest remains. Intermittent *meliy* gardens can progress in three ways:

1. If left to natural succession and not affected by invasive species or wildfires, these sites are likely to revert back to secondary vegetation and secondary forest (Cycle A).
2. In savannas or sites that have been cropped and burnt frequently until woody biomass has been exhausted and the soil impoverished, people must resort to more intensive ditched-bed gardening under full sun with only a grass fallow (Cycle B). Ditched bed gardening sites can revert to forest, as old ditched beds can be found below forest today.
3. If people settle in an area used for intermittent gardens, the fallow is managed to select and plant useful trees and the site is developed into an agroforest (Cycle C).

Should agroforests become weedy or threaten to shade out taro patches, the system may be reconditioned by killing less useful trees, severely pollarding useful ones, and converting the area into a *meliy* garden for a period of time. During the German and Japanese administration, some coastal areas were converted into coconut groves. Most of these areas have since become mixed agroforests or secondary vegetation. Increases in human populations tend to drive vegetation dynamics to the right while decreases in population result in shifts to the left (in Figure 4.8). Today these vegetation dynamics are increasingly affected by bulldozing activities, invasive species, wildfires, and climate change/ENSO events. Vegetation dynamics in wetlands are especially affected by sea level, storm surges, and road construction.

4.6 A Journey down the Watershed: Traditional Management and Modern Impacts

Figure 4.7 illustrates the general pattern of traditional land use in relation to watersheds. The sections below describe some traditional activities carried out at the upper, mid, and lower sections of the watersheds under study on Yap.

4.6.1 Traditional Management at the Upper Ends of the Watersheds

The watershed project resulted in the mapping of 1,959 sites where shifting *meliy* gardens have recently been made. Seven of these sites were located in the 1–2 m (3.3–6.6 ft) zone, 56 in the 3–5 m (9.8–16.4 ft) zone, and 1,896 in the 6–177 m (20–581 ft) zone. The great majority of these gardens are thus out of vulnerability range for salinization and represent important areas for future food security. As agriculture is moved upslope there will be a tendency to sacrifice the limited remnants of native forest in order to exploit their better soils, so maintaining forest biodiversity will require deliberate planning and investment in enhancing *meliy* gardening so that it will not be necessary to sacrifice native forests. Here again, it is instructive to consider traditional agriculture. The soils of upland areas where ditched bed systems and are still etched in the landscape today, consisting of heavy, generally waterlogged, nutrient-poor clay. Nonetheless, gardeners of the past were able to garden in such areas using techniques that may allow for the use of biochemical and microbial processes, as soils are moved from wet ditches and placed on drained garden beds (Falanruw 1995). This ditched bed technology is similar to some of the productive drained-bed and hydraulic agricultural systems of Latin America, and to systems used to grow sweet potatoes in Papua New Guinea (Denevan and Parsons 1967; Denevan 1970, 1982; Wilken 1987). Given the current focus on microbial agriculture, the potential of this approach within these systems bears investigation and investment for enhancement of Yap's upland agriculture.

4.6.2 Traditional Management within the Watershed: Streams and Taro Patches

The watershed project included support for related projects, and communities were invited to submit proposals for watershed projects within their villages. Three projects were submitted, which involved the renovation of traditional rock walls (Figure 4.9) to stabilize stream banks so that more upland taro patches could be renovated. Over generations, streams have been managed so that the right amount of gently flowing water was provided to taro patches without washing away the rich silts, which is needed to grow taro. These silts were carefully managed to grow Yap's main taro, *Cyrtosperma merkusii*, *lak*, a taro that can be integrated with agroforests to maximize production throughout the year. Narrow paths generally run along the sides of streams, serving as both levees and footpaths and are maintained along with the streams. This approach leaves room for human habitation and agroforest all within a limited area of (more level) village land. Increased rainfall and stream velocities have damaged rock walls, and the advent of wage employment for both men and women has reduced the labor force to maintain the upland portion of streams. Until recently, this was not a big problem as there are plenty of coastal taro patches to augment the situation. Unfortunately,

FIGURE 4.9
Section of stream stabilized by a hand-built rock wall. The rock wall prevents the stream from washing away silts needed to grow taro. Upland taro patches such as the one in the background are now being renovated to replace coastal taro patches that have been affected by saltwater intrusion.

high seas and more frequent storm surges are now affecting coastal taro patches and threatening food security. Thus, it is necessary to begin using upland taro patches more actively. Most of these upland taro patches have, however, been damaged as a result of erosion of stream banks during heavy rains. This has resulted in the loss of the silty soil from taro patches used to produce one of Yap's most important staple foods.

The streams associated with the agroecosystems described above are located in forested areas that can only be reached on foot, and work was done with hand tools. This project is already resulting in the protection of taro patches, which are again collecting the silts needed to grow taro and individual taro patch owners are now beginning to renovate their upland taro patches to replace those damaged by saltwater intrusion.

Unfortunately, sections of the repaired stream banks have been damaged as a result of heavy rains associated with a record number of typhoons passing near Yap during the project. Climate change predictions suggest an increase in storms and rainfall for Yap in the future, and it will not be feasible for villagers to continue to repair and maintain stream banks by traditional handywork with stones alone. The forestry staff working with the Yap watershed project considered practices available for stabilizing stream banks; most stream bank stabilization projects in larger countries involve more land and gentler slopes than are available in these villages, and the use of heavy equipment to move rocks, or the use of cement. The challenge is that these methods cannot be used without sac-rificing the taro patches, home sites, and agroforests that the project is try-ing to protect. Cement could be used but is difficult to transport up into the forest, and if a cemented rock wall is broken, it is not easily repaired. In addition, cement walls would remove habitat for stream life such as freshwater shrimp. The Yap watershed team thus consulted literature and a number of experts and determined that the best practice in this case would be the reinforcement of the rock walls with heavy mesh. Wire mesh has been used in several coastal areas of Yap with good results. Heavy mesh will hold rock walls in place and still allow percolation as well as accommodate habitat for stream life. Eventually tree roots will secure the mesh and strengthen the embankments. Funding is becoming available to obtain suitable mesh to augment traditional stonework in the project vil-lages and other locations on Yap.

4.6.3 Traditional Management at the Lower Ends of Watersheds: Taro Patches and the Interface of Fresh and Seawater

At the lower ends of watersheds where more water collects, drainage systems are widened to become taro patches. There are five main kinds of taro[2] grown in Yap: *lak* or giant swamp taro; *Cyrtosperma merkusii* (formerly *C. chamissonis*) generally referred to as "taro" on Yap; *mal* or true taro; *Colocasia esculenta*; *laiy*; *Xanthosoma sagittifolium*, an introduced taro sometimes called "Honolulu" though it actually originated in the Americas; and *Alocasia macorrhiza*, a hardy taro grown more in the outer islands than on mainland Yap. A fifth edible aroid, *Amorphophallus* sp. is rare and not known to be eaten today. Giant swamp taro, *C merkusii* and true taro *C. esculenta* are generally grown in taro patches. There are especially many varieties of C. esculenta, *mal*, which requires the most intensive culture and is most esteemed. There are also a number of varieties of the "giant swamp taro" *C. meruksii*, *lak*, which is currently the most important aroid crop in Yap. *X. sagittifolium*, *laiy*, and related species are generally grown in *meliy* or about the house. While not as esteemed, they are likely to become more prominent in the future when taro culture must be moved upland where there is less taro patch habitat.

4.6.3.1 Kinds of Taro Patches

A variety of taro patches are found throughout Yap from ditched bed areas in the savanna to secondary forest on slopes in valleys, near streams, and especially in tree garden agroforest village areas. While the general term for

---

[2] The term "taro" is used here as a general term for Yap's edible aroids. It more properly refers to *C. esculenta*, the common name for which is derived from its Polynesian and Melanesian names such as "taro" (Tahiti), "talo" (Samoa), "kalo" (Hawaii), and "ndalo" (Fiji). On Yap, however, the term "taro" has been applied to *C. meruksii* perhaps since TTPI days, while the Yapese term *mal* is used for *C. esculenta*.

FIGURE 2.1

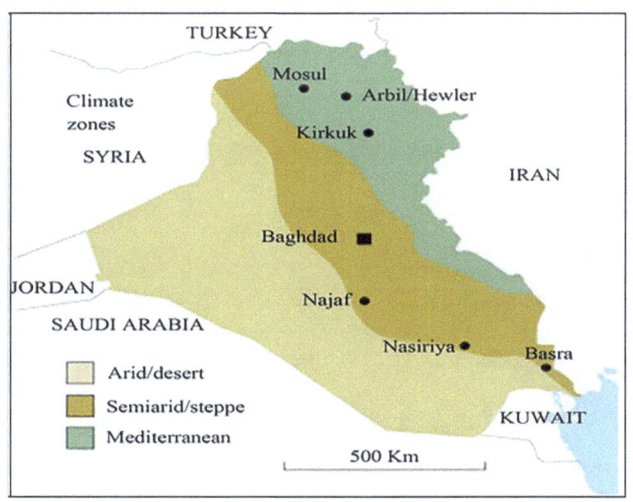

FIGURE 2.2

FIGURE 2.3

FIGURE 2.9

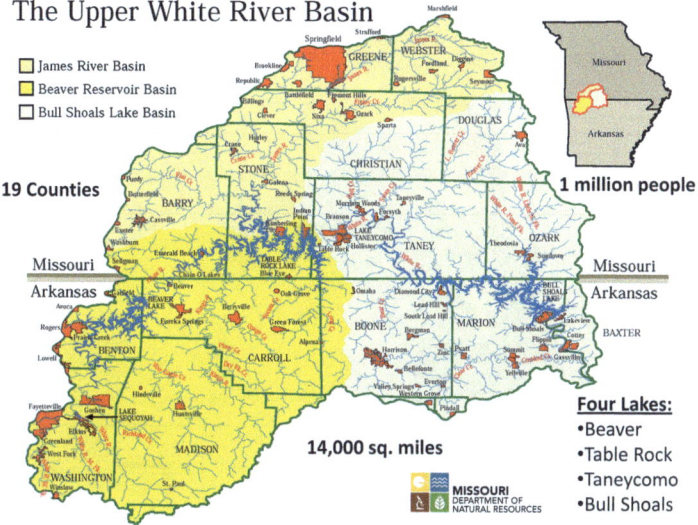

FIGURE 3.1

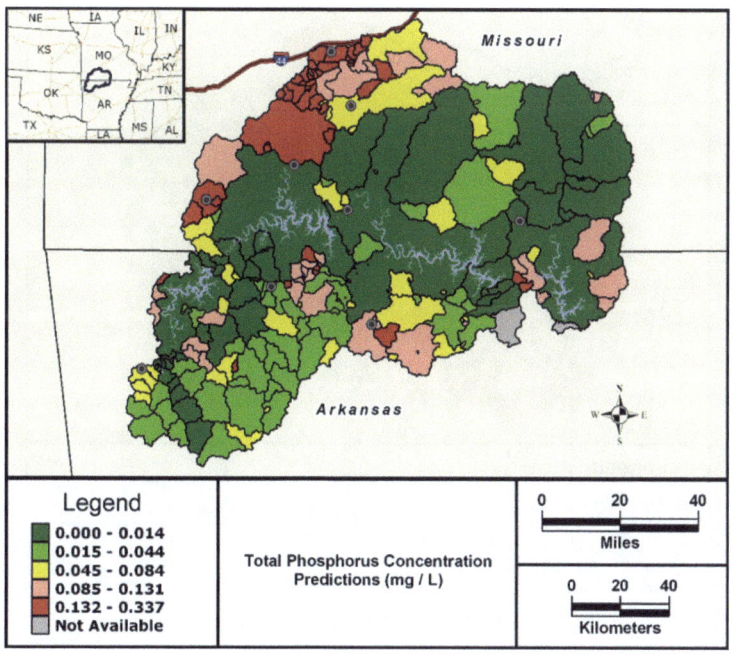

FIGURE 3.3

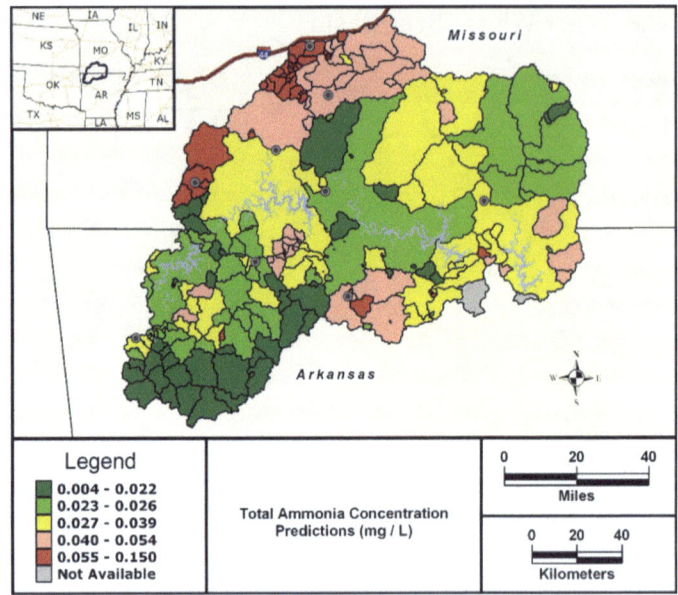

FIGURE 3.4

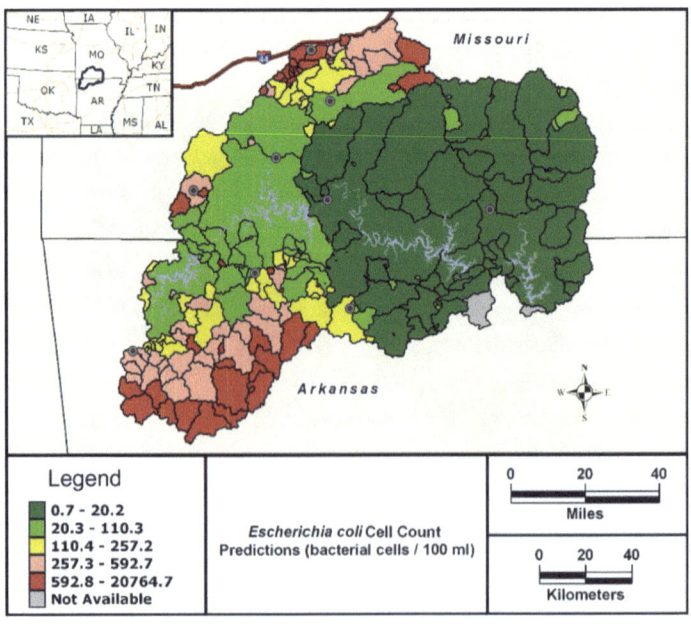

FIGURE 3.5

FIGURE 4.4

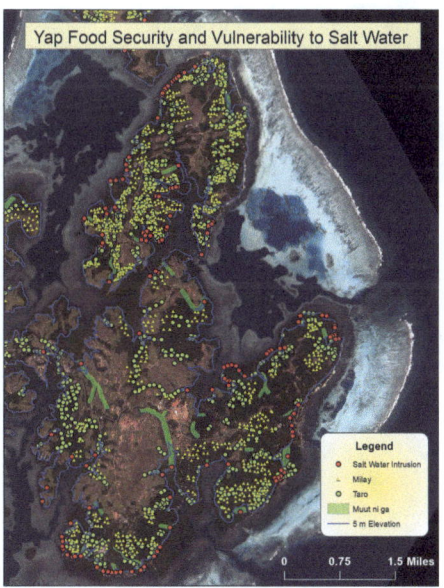

FIGURE 4.5

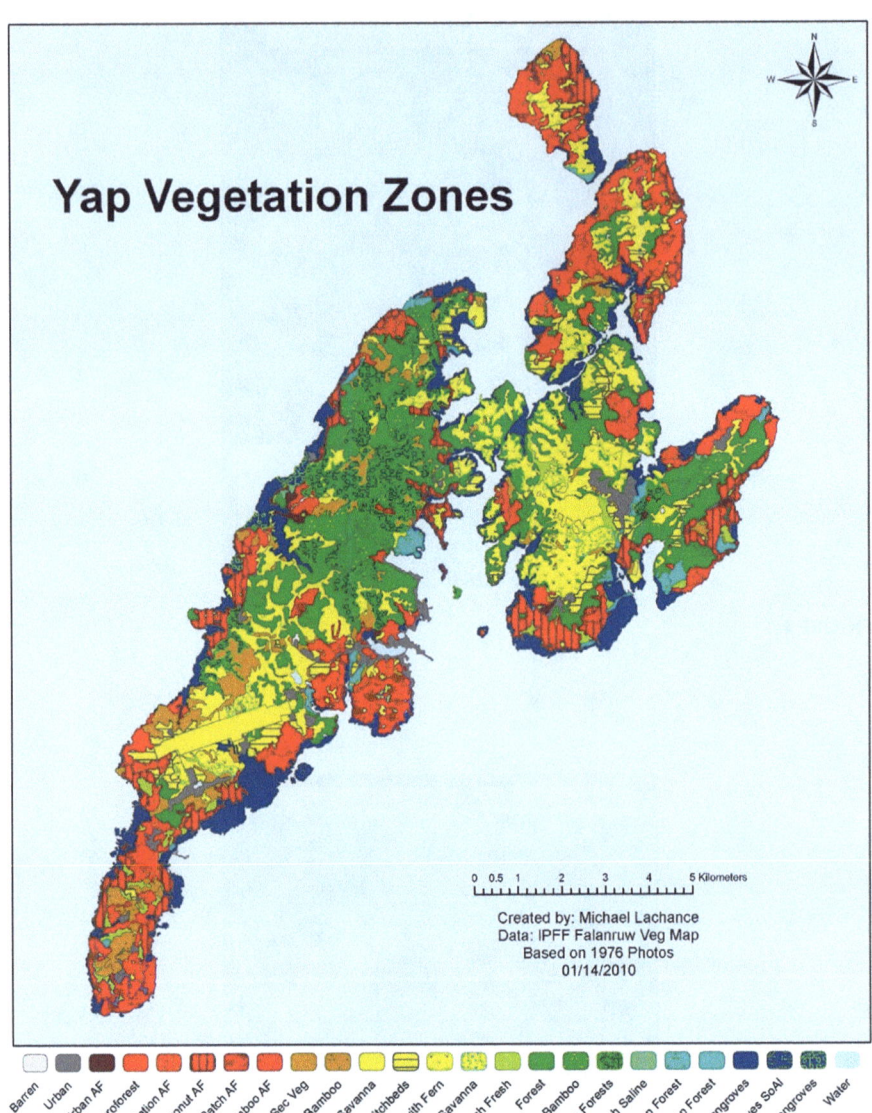

FIGURE 4.6

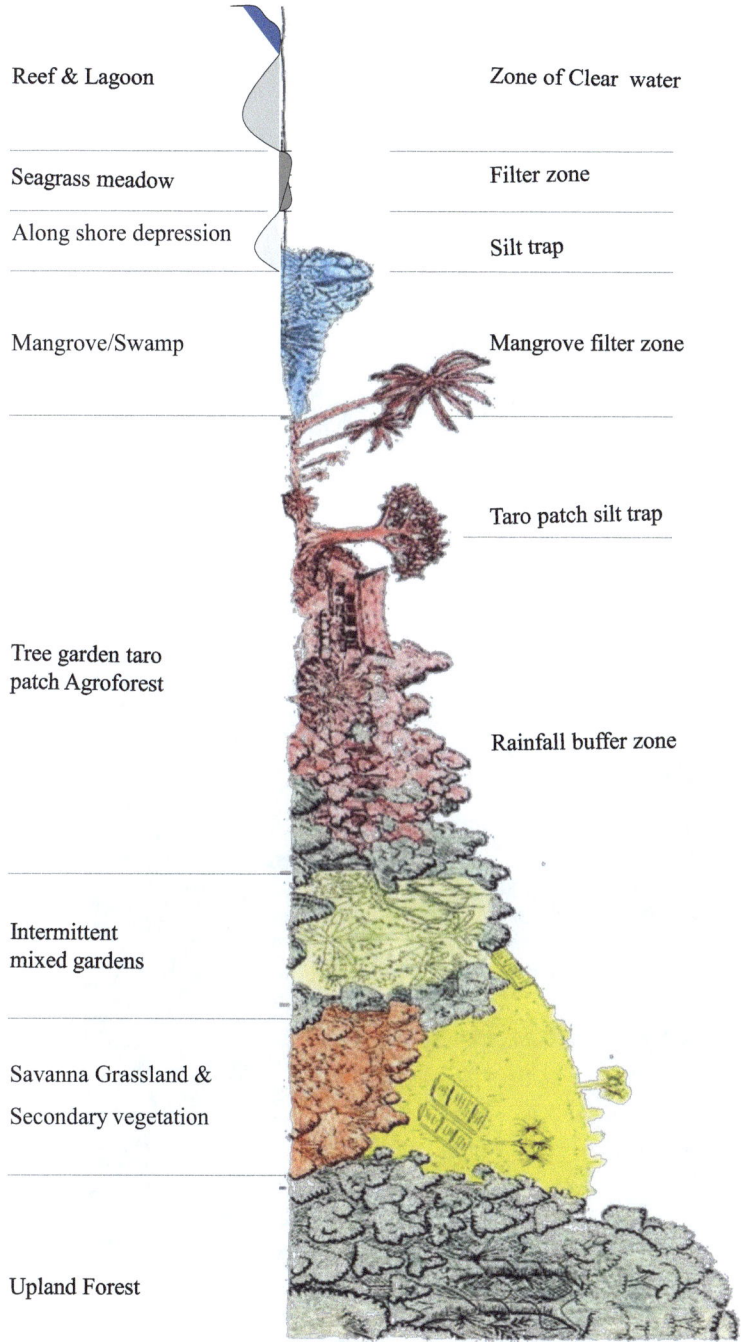

Reef & Lagoon — Zone of Clear water

Seagrass meadow — Filter zone

Along shore depression — Silt trap

Mangrove/Swamp — Mangrove filter zone

Taro patch silt trap

Tree garden taro patch Agroforest

Rainfall buffer zone

Intermittent mixed gardens

Savanna Grassland & Secondary vegetation

Upland Forest

FIGURE 4.7

FIGURE 4.11

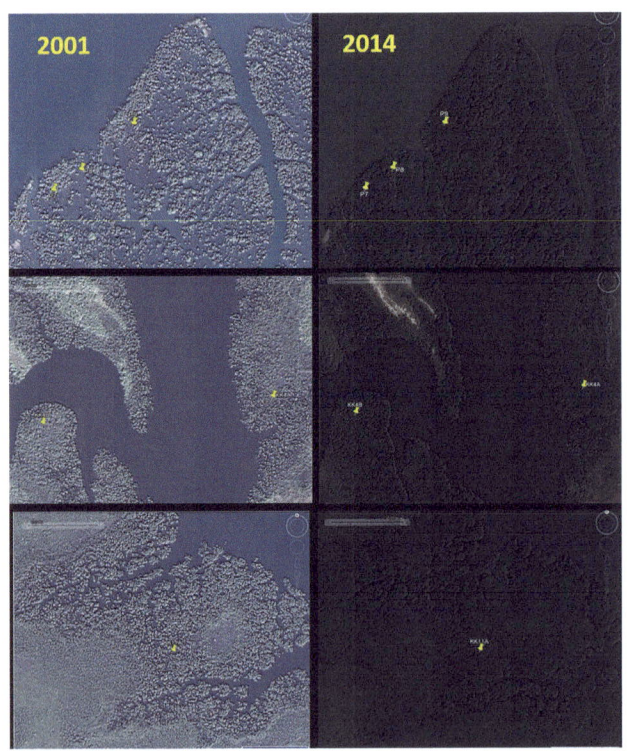

FIGURE 7.8

taro patches is *muquot* (*mo'ot*), there are a number of named categories based on where they are, how they were formed, and how they are used.

Taro patches found in the savanna and inland secondary forest are generally *talraq*, small taro patches created by enlarging ditches between garden beds. Most *muquot* taro patches occur in valleys, near streams, in agroforest, and in villages. Those that are formed when soil is removed from holes to create raised gardens are called *Milboquch* or *Wanuum* (in different parts of Yap). *Muqut ni gaa'* or *muqut ko Pumoon* are large taro patches formed in marsh areas that are naturally low. Some of these areas were formerly filled with *Hibscus tiliaceous*, the tree hibiscus that often grows in freshwater swampy areas. Traditionally most of the *Muqut ni gaa* were for men of the village though sections might be cared for by individuals. A small portion of a larger taro patch, or a small individual taro patch bordering other taro patches is referred to as a *goep*. Traditionally, *muqut ko Pumoon* were tended by men or older women for men, and young women did not work in these taro patches. Some taro patches were for families and they were traditionally tended by women.

4.6.3.2 *Moisture and Water Management in Taro Patches*

Both *Colocasia* and *Cyrtosperma* taro can be grown in either garden areas or taro patches. Some *Colocasia* is grown in moist areas of gardens, and in the past *Cyrtosperma* was sometimes grown in ditched bed garden areas. It is most productive, however, to grow these taros in taro patches where there is constant moisture. During droughts, soil is mounded about taro plants to prevent them from drying out and to prevent the suckers from detaching prematurely, and ditches may be dug to direct water to the taro plants. Some tree roots, such as those of *Hibiscus tiliaceous, Barringtonia racemosa,* and betelnut, invade taro patches and tap the water. These trees are cut back to avoid competition of their roots for water and nutrients. Some may be left however to provide shade to "encourage" taro to grow tall.

Traditionally many methods were used to manage water flow in relation to taro patches. Stream banks were lined with stone to prevent water from flowing into taro patches or to remove excess water away from taro patches. Water flow was managed by banks called *n'ey*, or by varying the width of a ditch or stream. Sometimes, ditches were dug into slopes above taro patches to drain away excess water and allow it to gently seep through the soil into taro patches. In such cases, the bottom of the ditch was kept somewhat higher than the taro patch so as to maintain an appropriate water level within the taro patch. It was also necessary to manage the depth of streams. If drainage ditches or streams became too deep, they would drain too much water from the surrounding area and tend to dry up taro patches. In such cases, berms were built to trap soil and small rocks so that the bottom of the stream was raised. Another ingenious method of raising the level of a wide ditch or stream was to plant closely spaced betelnut palms along the stream bank.

The roots of the betelnut palms served not only to hold the banks, but they also develop adventitious rootlets that grow out and up into the stream. These fine but strong roots serve to filter out rocks and other debris so that the bottom of the stream bed is raised, thus raising the water table in the area to a level conducive to taro growth. Should the stream become too shallow, the betelnut palms can be cut down.

Most taro patches have an inlet and outlet for water to flow through the taro patch. Some small individually dug taro patches may have only an outlet to drain water after heavy rains. The inlets and outlets, called *wun*, are managed to ensure appropriate drainage and a gentle flow of water so that the taro patch does not get stagnant or too shallow or too deep. The velocity of water flow in drainage ditches was managed in order to prevent erosion. This was done in a number of ways including managing the slope of the ditch, the creation of deeper holes within the ditch, and the lining of ditch bottoms with rocks. During periods of drought, the outlet channel of the taro patch might be blocked with soil or with rocks wrapped in the sheath of a betelnut inflorescence (i.e., *wathir*).

If a taro patch area is too deep, a raised bed may be built with materials on the site such as logs, coconut husks, and fern roots and filled with soil. This provides a planting area for taro so that it won't drown in the deep water. Another method used for growing taro in deeper areas involved the weaving of baskets of coconut fronds and then filling them with good soil to form an elevated growing "pot." In some deeper areas, the baskets were elevated and held in places with sticks. Taro is planted in the raised baskets, and when it becomes large, its roots are able to grow through the bottom of the basket and make use of the fresh water and nutrients below. An extreme case of such "taro hydroponics" were the floating taro patches developed in a few places with very deep fresh water marshes. In these Yapese taro hydroponic systems, taro was planted in a floating mat of plant roots and debris (Falanruw 1992).

During the project, Yap experienced a severe drought. The drought followed an intense ENSO "El Nino" event as was the pattern with the especially severe droughts of 1982–1983 and 1997–1998. The drought of 2015–2016 resulted in the driest period from October to March on record as well as more limited rainfall in the months before and after this period. As a result of the drought many taro patches, especially smaller man-made taro patches, such as *milbuquch*, dried up. Many taro patch owners harvested taro and took advantage of the dry conditions to recondition their taro patches. Unfortunately, the dry conditions killed many of the remaining corms that were needed to replant the taro patches. In the past this planting material might have been obtained from the deeper areas of *muquot ni ga*. Unfortunately, over the years of abundant rainfall associated with a prolonged "La Nina," many of these deeper taro patches have not been tended and have become overgrown by tall *Phragmites karka* reeds that shade out the taro. Thus, a lesson to be learned from the drought of 2015–2016 is that the old people had a reason for maintaining even those taro patches that were located in deeper areas.

4.6.3.3 Management of Saltwater Inundation and Intrusion

The development of raised taro beds and "basket culture" also serves to place the roots of the taro above saline water so that the taro is not as affected when these taro patches, many of which are coastal, are temporarily exposed to salt water. This method was used in some low sandy areas that had been affected by salt water after Typhoon Sudal, but in most cases, the taro was killed by subsequent and more prolonged saltwater intrusion.

Especially high "king tides" or storm surges can fill coastal taro patches with salt water and kill the taro. After such events, taro plants are harvested before they go bad. In gently sloping coastal areas, drainage ditches serve to drain salt water after especially high tides and storm surges and subsequent rains flush out the salt water. After Typhoon Sudal in 2004, many taro patches were renovated by the clearing and repairing of the drainage systems. Post-typhoon rains flushed out salt water so that taro could be replanted, sometimes growing better than before.

After Typhoon Sudal, one village with limited coastal area suitable for taro rehabilitated a taro patch that was subject to flooding after heavy rains, and also subject to flooding with seawater by very high tides. A dike was constructed along the lower part of the taro patch to keep salt water out at high tide. In addition, a pipe with a 90° angle was installed through the dike. When the taro patch was affected by heavy rains, the pipe was lowered at low tide to enable the taro patch to drain. At high tide, the pipe was raised so that salt water could not flow into the taro patch.

Sometimes, in more level coastal areas, dikes alone cannot prevent saltwater intrusion from below as the pressure of a high tide enables salt water to percolate through the porous soil into the taro patch from below. In some such coastal taro patches, a deep trench was dug on the landward side of a dike or stone path parallel to the coast. This trench was filled with stones and allowed to fill with fresh water. The pressure of the fresh water then served to reduce saltwater intrusion from below, during high tides.

4.6.3.4 The Use of Diversion Swales and Check Dams

Relative to today, more macro management was traditionally applied to taro patches that occurred in marshy areas of valleys that emptied into mangroves, and the sea. In such places water flow was diverted at the head of the valley into wide shallow swales that ran along the side slopes of the valley parallel to the taro patch. This prevented water that flowed down from upslope areas from rushing into the taro patch and washing out taro plants and eroding away the sediments needed to grow the taro. At the same time the swales allowed for a gentle percolation of water through the soil into the taro patch so that it remained moist. Roots, often of *Hibiscus tiliaceous* trees, prevented the erosion of the soil into the swales. Running cross-slope and parallel to the taro patch, the swales also provided a convenient walking path for reaching taro patches. As people utilized the path they also maintained it, cutting away intruding vegetation or shoring up side banks so that they were not eroded away.

The management of water runoff in swales across valley slopes was complemented by the development of low check dams at the lower end of the taro patch. These low dams served to further slow the flow of water out of the taro patch so that sediment was deposited, leaving a thick layer of nutrient-rich soil for taro. At the same time, the dam prevented the flow of salt water into the taro patch. The height of the dam was adjusted so that it allowed excess water to drain from the taro patch after periods of heavy rain, but high enough to prevent the entry of seawater at high tide. From time to time, the dam was adjusted to adapt to conditions such as extra high king tides.

4.7 Contemporary Impacts on Traditional Agriculture

Few of the diversion swales described previously survived the Japanese occupation of Yap and World War II, as Japanese agriculturalists disrupted the system so that the taro patch would fill with soil and silt and could be used to grow other crops for the large population of Japanese soldiers occupying Yap prior to the war. By the end of the Japanese administration of Yap, there were only about half as many Yapese and not enough labor remaining to restore the original traditional agricultural system. In its place, Yapese women took to managing the water in taro patches with ditches and berms along and within the taro patches. These relatively "micro-managed" agricultural systems continue today and only remnants of the former system of diversion swales on valley slopes remain.

The system of check dams at the lower end of taro patches has also fallen into disuse, resulting in major problems. Without the dam, silt from the taro patch is eroded away and it travels from where it is needed as a growing medium into the marine environment where it becomes a pollutant.

As the soil erodes away, the drainage channels within the taro patch become deeper. As the discharge velocity increases, stream banks are eroded in a process of bank caving. As the sides and bottom of the channel erodes further inland, drainage from the taro patch is increased and some areas are left too dry for the cultivation of taro. Without the presence of the dams to maintain a pool of fresh water upslope, salt water is able to flow further into the taro patch and to bubble up into the taro patch at high tide. This results in saltwater intrusion reported in many places in Yap and especially in the outer islands of Yap.

The problem of progressive bank caving is occurring in many areas on Yap, drying up taro patches and allowing saltwater intrusion. Climate change will increase this problem as predicted increases in heavy rainfall result in increased erosion and sea-level rise results in increased saltwater intrusion. Despite this predicted scenario, the reapplication of traditional technology can be used to extend the life of taro patches both on Yap and in other areas of the Pacific with coastal taro culture.

The labor force available to restore taro patches and to optimally manage other traditional agricultural activities is not adequate to achieve the large task at hand. While the population of Yap has risen rapidly during the American administration of the island, this has meant a larger proportion of children to mature women. In addition, the opportunity for women to attend school and to obtain paid jobs has kept the agricultural labor force relatively low as women are the main agriculturalists. The depletion of secondary forest resources and problems of invasive species have made *meliy* gardening more problematic and the number of more proficient gardeners and taro patch stewards has declined due to age and recruitment as baby sitters. More recently, changes in the day to day pattern of rainfall have impeded controlled burning activities to initiate *meliy* gardens.

Taro patches have suffered from a general levelling of land as former landscape architecture of ditching and raising or lowering areas has not been maintained, and the labor force has not been able to manage the flow of water in rivers resulting in erosion of silt or filling in of taro patches. In addition, storm surges and king tides have carried salt water into taro patches lacking adequate drainage ditches to drain saltwater inundation. In the aftermath of typhoon Sudal in 2004, a program to support the restoration of taro patches became available, enabling villagers, mostly women, to renovate many taro patches. Aspects of the program were inspiring as some older women practically crawled to sites to instruct younger women on what needs to be done to restore taro patches. The program was successful in restoring enough taro patches that United States Department of Agriculture (USDA) disaster relief supplies of rice were discontinued much earlier than in the aftermath of typhoons on other islands.

Coastal road construction has greatly impacted taro patches by preventing the drainage of fresh water or, in cases where there has been an inundation of salt water from a storm, the road prevents the drainage of the salt water. In some such cases a salinity gradient may develop in taro patches such as the one shown in Figure 4.10. Conductivity measurements within this taro patch show that the surface waters are fresh but water becomes increasingly saline with depth. Many deeper taro patches are like this one, with giant swamp taro, *lak, Cyrtosperma meruksii* still growing in shallow fresher surface water around the perimeter and water spinach, "kangkong," *Ipomoea*

FIGURE 4.10
Salinized taro patch with *lak* taro, *Cyrtosperma meruksii,* growing around perimeter and water spinach, *kangkong, Ipomoea aquatica,* growing in surface waters at the middle. Gregory Chem holds a pipe that will be sunk into the taro patch to serve as a monitoring station. Taro in the background is yellowing due to salinization from a recent storm surge.

aquatica, growing in the center. The water spinach requires fresh water and can grow in the taro patch because it floats at the surface where the water is fresh. However, taro can no longer grow in the saline soil in deeper parts of the taro patch.

4.8 Sustaining Biodiversity and Ecosystem Services

The Micronesia Challenge effort upholds that the MC "is a commitment by the Federated States of Micronesia, the Republic of the Marshall Islands, the Republic of Palau, Guam, and the Commonwealth of the Northern Marianas Islands to preserve the natural resources that are crucial to the survival of Pacific traditions, cultures and livelihoods. The overall goal of the Challenge is to effectively conserve at least 30% of the near-shore marine resources and 20% of the terrestrial resources across Micronesia by 2020" (www.micronesiachallenge.org; checked September 15, 2018). In the FSM, the criteria and designation of terrestrial areas are left to each state.

In terms of conservation of native species, the most important habitat type is native upland forest; however, Yap does not have 20% relatively undisturbed upland forest. Due to the conversion of most wild forest to *meliy* gardens at some time in the past, most wild forest is largely secondary. Therefore, candidates for the bulk of Yap's portfolio of terrestrial protected areas would need to include three forest types (Yap State 2014). These include all remaining remnants of better developed native upland and swamp forest, mangroves, and better developed agroforests.

Mangroves make up about 12% of Yap and are both native forest and keystone habitat important for coastal protection, fisheries, and bird habitat. In addition, research on Yap (Donato et al. 2011, 2012) indicates that while mangroves make up 12% of the islands, they sequester 32% of the carbon that is removed from the atmosphere by Yap's vegetation. They are, thus, important in reducing levels of greenhouse gases contributing to global warming and sea-level rise and may have added economic value as carbon markets develop.

Yap's agroforests make up about 26% of the islands area. They are the most diverse and well developed in Micronesia, incorporating not only high agrobiodiversity but also landscape-architectural elements. Being important for both food security and ecosystem services, they are likely to make up a significant percentage of Yap's portfolio of terrestrial protected areas. While representing the achievements of the past, Yap's agroforests represent potential for sustainable futures. The limitations of Western agriculture in terms of sustaining biodiversity and ecosystem integrity are being realized (United Nations Environment Programme 2009), and the value of agroforests for conservation of biodiversity in tropical agricultural landscapes is being realized (Gotz et al. 2004). There is potential for convergence of Yap's traditional practices of agroecology with the evolving science of ecological agriculture (McNeely and Scherr 2002; Beck 2013), and farming in harmony with natural systems (McNeely and Scherr 2007). This could make Yap one of the most modern nations, as the world's economic systems become better aligned with ecological systems (Daily and Ellison 2002), and feature Yap as a leader on ecologically sustainable development that must emerge from the current changing climate crises.

Finally, while they are very small in area, the less impacted uninhabited islets in Yap's outer islands are locally and globally important as sea turtle and sea bird nesting areas, and as habitat for coconut crabs, endemic fruit bats, and lizards, as well as a recently described endemic species of snake (Wynn et al. 2012). The few islets with inland mangroves are important as gauges of sea-level rise. An example of an evaluation of five uninhabited islets can be found in the work of Falanruw and Ruegorong (2010).

4.9 The Importance of Community Action

In the introduction to the FSM NBSAP (2002), the Secretary of the FSM Department of Economic Affairs acknowledges, "Most important are the people in the communities and their leaders who, on their own time and through their concern for the welfare of the biological resources and the needs of the next generation, participated in and provided their views at community workshops. They are the most valuable resource in this whole process. It is on their commitment that we must base our dedication in ensuring that we achieve the vision by effectively carrying out all the actions outlined in this Plan for our sake and that of the future generations."

Natural resources cannot be managed without involvement of the people who own and use these resources. Almost all of the natural resources on Yap are held under private customary tenure and there are no examples of successful government protection of biodiversity. Indeed, there are no conservation officers on Yap. Large old growth trees are being depleted by introduced sawmills and there is no enforcement of laws prohibiting the commercial exploitation of vulnerable species such as native pigeons, sea turtles, coconut crabs, napoleon wrasse, bump head parrot fish, and other reef fish or spawning aggregations of groupers. While customary practices and prohibitions, resource apportionment, and customary land tenure helped to avoid the tragedy of the commons (Hardin 1968), which involves "ways of the past," the way in which resources are used contemporarily on Yap is becoming more of an individual choice, rather than being exploited by commercial interests outside of the culture. Thus, management efforts aligned with the modern context are needed.

Many meetings relating to sustainable management of natural resources have recommended community management and government–community partnerships, and these approaches are emerging. Examples on Yap include the Nimpol Marine Protected Area, the Tomil Resource Conservation Trust, and the recent Weloy Resources Plan that establishes the Weloy protected forest. The GIS database developed by the watershed project, described in this chapter, is available to all communities on Yap to assist with management planning and to develop data-based proposals for support from a number of sources. These include the ongoing U.S. Forest Service and other grant programs that are becoming available in recognition of the importance of community action. Regulations for a Protected Area Network are being developed. These will link government and community efforts and enable communities to obtain support from the MC Trust fund.

The Weloy Resources Stewardship Plan resulted from cooperative efforts of the community, the Yap Community Action Program, a locally based Yapese staff of TNC and Yap Forestry, with technical assistance from the USFS IPIF and YINS. Continuing efforts include a socioeconomic component for the project and additional forest assessments assisted by drone imagery, provided by the QU/Yap program. Figure 4.11 is a composite of drone imagery of the Weloy Forest Reserve with insets showing community members pausing under a heritage tree during a survey of their forest and community leaders gathered to sign the Municipal plan, which established the forest reserve.

Much work remains to be done. A detailed documentation of Yap's agrobiodiversity is needed. An assessment of salinity in taro patches is ongoing with assistance from community NGOs and village residents. It is

FIGURE 4.11
(See color insert) Drone imagery of the Weloy Forest Reserve with insets of community gathering to sign Weloy Resources Stewardship Plan (inset at the top), and (inset at the bottom) community members pausing under a large *biyuuch, Calophyllum inophyllum*, heritage tree during a survey of their forest. The drone imagery shows the mosaic nature of the forest which also contains cultural and historical sites, and is the location of a number of rare endemic species. The light-colored open area on the right of the main image was cleared for a wind turbine farm prior to the declaration of the protected forest.

anticipated that results from this project will inform efforts to better manage the interface of fresh and salt water when coastal roads are developed in order to protect both coastal taro patches and also mangroves needed for coastal protection and carbon sequestration. Support is becoming available to obtain mesh materials to augment the strength of traditional stonework along streams while still allowing for habitats for stream life and the growth of root systems to stabilize streams running through agroforests. This will address predictions of increased rainfall and enable the renovation of more upland taro patches to replace lower elevation taro patches affected by saltwater intrusion. LiDAR imagery is being sought to enable a more refined assessment of coastal taro patches as well as a better understanding of traditional upland water management systems than is possible with the current QU/Yap DEM, which is based on 5m intervals of the existing USGS map. Food production systems are moved upland, yet research and applications of biointensive and biochar systems are needed to enhance traditional ditched bed and *meliy* systems to decrease the need to sacrifice limited remnants of native forest.

Maintenance of the Earth's ecological system is emerging as one of the greatest challenges facing human society (Lubchenco et al. 1991), and modern agriculture is faced with the challenge of producing economically viable crops, while preserving the short- and long-term integrity of local, regional, and global environments. The need for more sustainable agriculture is becoming a national and global priority (National Research Council 1989). It is useful to look at systems of food production that do not require large subsidies of chemicals, fossil fuel energy, and heavy machinery. Likewise, with growing concern about global climate change and sea-level rise associated with increased levels of carbon dioxide and other greenhouse gases, there is a need for consideration and support for more nature-integrated technologies of food production, such as those found on Yap.

Disclaimer

The research, analysis, and other work documented in this chapter was fully or partially funded by the USDA Forest Service; however the findings, conclusions, and views expressed are those of the authors and do not necessarily represent the views of the USDA Forest Service.

5

Halal Food Security: Forensic and Laboratory Management

Irwandi Jaswir, Muhammad Elwathig S. Mirghani, and Bubaker Hamad Almansori

International Islamic University Malaysia (IIUM)

Fitri Octavianti

Universiti Sains Islam Malaysia (USIM)

CONTENTS

5.1 Overview of the Societal Imperative

Currently, with the advancement of science and technology, food processing has become more complex, and foods may contain ingredients and additives from questionable sources. With efficient transportation more new products are introduced onto the shelves in the market across the world and somehow, somewhere foods can be lost in the food chain or not traceable at all. It becomes a crisis when a food causes hazards to humans and a food product is recalled, or a producer is summoned to curb distribution and stop its consumption. In this case, the assurance of authenticity is a primary concern so as to address the problems at the source, and to ascertain the production unit(s) responsible. Such are the precepts of food security, which is a fundamental necessity of life that all humans share, without which people can experience harm to their health and well-being.

The need for proper control and monitoring of the authenticity of food is a serious matter to the authorities and food manufacturers, and a strong commitment and continuous support from the government through various agencies can ensure the integrity of the food supply itself, both in terms of safety and quality. People of certain faiths must also have confidence that they are consuming food and drink of vegetable origin, such as in Islam, where all things created by God (Allah) are permitted (halal) to be eaten, with a few exceptions that are specifically prohibited (haram), which are pork, blood, carrion, and animals slaughtered without invoking the name of Allah. Islamic food laws are based on cleanliness, sanitation, and purity. All utensils must be clean and free of contamination from any religiously prohibited or harmful substances, such as is determined in Islam. Islam does not prohibit vegetable foods, with the exception of what is fermented, whether it be grapes, dates, barley, or any other substances; as long as it remains in the unfermented state these items are acceptable. Similarly, Islam prohibits anything which intoxicates, affects the functioning of the brain, or harms the body (Al-Qaradawi 1984). However, with regard to foods derived from animal sources, people and nations have held widely varying attitudes about these foods. Hence, the importance of establishing laboratories and using analytical techniques (methods) of validating authenticity in halal food production for ensuring food safety and protecting consumers from fraud and deception is necessary, as well as for product recall purposes, with regard to halal foods. Halal analytical laboratory facilities should fit these purposes and ensure competent personnel and equipment. Laboratory data may help define the overall scope of work, levels of worker protection, as well as remediation and disposal methods. The laboratory information management system should also be stated in such a way that allows matching analytical results with relevant field data. Instrumental methods in detection of contamination and/or adulterants in food help to clarify any doubt to Muslims about the quality of foods, and this information can also be disseminated for consumer transparency through labeling, providing better trust and higher confidence from the consumer perspective. To accomplish this goal, laboratory staff working with halal foods should include chemists, biologists, engineers, chemical engineers, laboratory technicians, laboratory assistants, as well as other supporting staff for administration, accounting, security, and cleaning. All staff should be trained to develop the laboratory's leadership skills to improve customer satisfaction, safety, employee morale, and even the bottom line of food production and distribution. Engineers, laboratory technicians, and assistants should do continuous quality checks and calibration of instruments in the laboratory as well as annual maintenance checks by the specialists or by suppliers responsible for delivering the products, when the line of production is bound by the rules and regulations using halal ingredients for halal production. To this end, much laboratory analysis is needed when necessary, or if any doubt about processing or contamination of a specific batch or product occurs.

Exchange of technical expertise in laboratory analysis of halal food and sharing the information and data will help communities of practice in the food industry to enhance the capacity for halal food product development. This work, which originates in Malaysia, helps to boost the image of Malaysia and impacts many Muslims outside of Malaysia, at the international level, specifically in terms of food security.

5.2 Global Societal Impacts

In modern society, a shift in our traditional approaches, compared to our ancestors, has resulted in a life where we eat and drink as consumers, no longer consuming foods and beverages that are directly and personally produced and processed, and instead we are dominantly dependent upon intermediaries in the food supply chain for consumables. In this process, many consumers may not know (or may not need to know) how the food processing was carried out and how those intermediaries are involved in the supply chain. Nevertheless, Islam has laid down a code of conduct and rules to maintain a balance between the process and requirements of humans and their environment. These are the traditions that have gone unchanged in this modern set of changing circumstances. Islam allows humans to utilize the given bountiful resources to fulfill the necessities and circumstances of day-to-day life, including food, and yet we must maintain religious traditions. Religion plays a very influential role in food choice (Dindyal and Dindyal 2003), which is the focus of this chapter's description of intensive scientific work. The backdrop and driving force of this scientific work is the primary guiding influence of Islam, specifically on food consumption, which depends on the extent to which individuals follow the teachings of their religion (Abdul et al. 2009). For all of these reasons and the need to meet such primary needs of an extremely large "global community," namely, Muslims of the world, halalness is a primary additional need for food quality. Therefore, the detailed approach taken by the systems investigated in this chapter directly serves, currently, approximately 24% of the world's population (i.e., 1.8 billion people who practise Islam).

Halal food legislation, quality controls, inspection laboratories, education, and training are the lynchpins of a global system of maintaining halalness of food supplies, and an integration of these system components applies directly to the practice of food forensics and laboratory management, such as laboratory operations, staffing and assignments, laboratory relations, and high-quality leadership. The systematic approach to maintaining halalness of the food supply also mitigates the presence of risk and uncertainty when making purchasing considerations, and the establishment of highly professional halal laboratory management is very important to instilling with Muslim consumers the confidence that food to be purchased is halal.

To improve success and thus reduce risk to the consumer, we have developed an audit risk framework in conventional assurance services, which may help to improve the success of a scientific approach to halalness in a business environment, allowing for increased value and credibility of halal products that display a halal logo. Gradually, the building of consumer trust helps to build the perception of decisions concerning complaints or other disputes that arise about producers and associated companies (Santos and Fernandes 2008; Abdul et al. 2009), and so a science approach to the problem of establishing known halalness, provided by this work, serves some positive characteristics, such as trust and loyalty, furthering the establishment of trusted relationships between customers and companies, thus building societal bonds that are very important to contemporary society. In this way, we continue the religious traditions of Islam in a contemporary, and increasingly consumer-driven, global society. Accordingly, Hermann (1997) and Ahmed (2008) describe the importance of the first stages of purchases, which take place when the consumer first inspects the food they are considering consuming, followed by requests for additional information from the vendor's presentation. This means that an effective label that expresses the compliance of food with the information that satisfies the fundamental values of the consumer (i.e., halalness) might be a better cue for quality than other cues, ultimately leading to a purchase, consumption, and satisfaction. Quality management practices such as adopting halal certification can also boost customers' confidence and, hence, can lead to their satisfaction. Customer satisfaction is, intuitively, both historically and contemporarily, a fundamental human need, which is addressed by this scientific work.

Indeed, business is a powerful and integral element of contemporary society, and thus, the business-oriented focus of the work in this chapter has major global societal impacts, and it is for this reason that the environmental sciences (i.e., agricultural, bioscience, and food sciences) conducted in this chapter strongly influence positive relationships between food producers and consumers, in terms of quality and customer satisfaction and business profitability (Anderson et al. 1994). Also, it is the work of science that fosters positive psychological features, which have a great effect on halal purchasing power, depending on behavioral descriptors of the Muslims in a certain local community, country, or region of the world. In this way, the environmental science work described in this chapter has a positive impact on maintaining and building upon the traditions of the Muslim World, thereby supporting important dimensions of our contemporary global society.

5.3 Sampling: Connecting Food Sources with the Consumer

The most important part of a halal analysis, or any type of specimens to be analyzed in a laboratory, is the sampling. Sampling techniques must ensure that the items tested are truly representative of the stock from where they were taken. Sampling should be done by an expert who is well trained or with personnel that are well supervised. Sampling examples need to include codes with a serial number, date, and time of processing, and special rooms for receiving, preparing, and numbering samples should be established. All results should be reported on standard forms that have all of the information collected about the sample, including the type of analysis, results of analysis, special comments, and official signatures and stamps of certification. All of these steps are the important elements of registration of the services and supplies and can be very helpful for the assessment of fitness for the purposes intended, and for any instances where a trace-back is necessary.

5.4 Challenges in Halal Food Analysis

Halal food in the contemporary food industry is food of high quality and safety, which conforms to international standards of food safety according to "Hazard Analysis and Critical Control Points," and it should be permitted under the Islamic Shariah law. It is very challenging and increasingly difficult for Muslims to ensure the halal status of food in the market due to the diversification of sources acquired globally for food processing and production. This trend has raised concerns among Muslim consumers regarding processed food. Adulteration of value-added food products involving the replacement of high cost ingredients with lower grade and cheaper substitutes can be very attractive and lucrative for food manufacturers or raw material suppliers. Many fraudulent cases reported worldwide involving adulteration of haram ingredients in halal food (especially porcine-based products) has occurred. In other cases, non-halal contaminants were unintentionally introduced in the final food products.

Muslims across the world acknowledge and value the importance of living a healthful lifestyle, and complying with their religious beliefs. There is a need for high confidence in the appropriateness and purity of the food they consume. For example, Islamic teachings prohibit the consumption of blood of any type and consider it haram. Halal food is a sensitive and serious matter to Muslims. With many fraudulent issues occurring and cases of unintentional non-halal contaminants in food, we recommend that more stringent monitoring be established by competent certification authorities. Authentication and verification for halalness has become one of the major challenges in analysis of processed food. At present, very limited analytical methods are available for halal food verification. Rapid, sensitive, reliable, and yet affordable methods are urgently needed for halal food verification and for detection of non-halal components (e.g., porcine origin) in food products.

5.5 Current Analytical Methods That Can Be Used for Halal Food Authentication

For any type of food analysis in laboratory conditions, analytical techniques should be well established, documented, and clearly written and posted using the same protocol and instruments so that among different scientists there will be continuity and comparability of results from the analyses. Taking the approach of high-quality laboratory standards and a variety of current analytical methodologies to choose from, outlined below, allows for a variety of applications of technologies to maintaining halalness standards for food.

5.5.1 Gas Chromatography

Gas–liquid chromatography (GLC), or simply gas chromatography (GC), is a common type of chromatography used in organic chemistry for separating and analyzing compounds that can be vaporized without decomposition. Typical uses of GC are for the determinations of non-halal ingredients in food or for the analysis of toxicity, which make the food non-Toyyib, i.e., non-halal.

To be suitable for GC analysis, a compound must have sufficient volatility and thermal stability. If all or some of compound's molecules are in the gas or vapor phase at 400°C–450°C or below, and they do not decompose at these temperatures, GC can probably be used to analyze the compound. Analysis of foods in these circumstances is concerned with the assay of lipids, proteins, carbohydrates, preservatives, flavors, colorants, and texture modifiers, as well as vitamins, steroids, drugs, pesticide residues, trace elements, and toxins. Most of these components are nonvolatile, and although high-performance liquid chromatography (HPLC) is now used routinely for much of the food analysis around the world, GC is still frequently used. As an example, derivatization of lipids and fatty acids to their methyl esters (FAMEs) or proteins by acid hydrolysis, followed by esterification (*N*-propyl esters), and of carbohydrates by silylation, produces volatile samples suitable for GC analysis. Accordingly, GC analysis is used to detect the fatty acid composition. We found that lard differed with cow fat in C20:0, C16:1, C18:3, and C20:1, and with chicken fat in C12:0, C18:3, C20:0, and C20:1 fatty acids (DeMan 1999). Lard and chicken fats are significantly different in the disaturated and triunsaturated triacylglycerols. Marikkar et al. (2002) have used GC to determine fatty acid composition of refined–bleached–deodorized (RBD) palm oil and a series of RBD palm oil samples adulterated with enzymatically randomized lard. A gradual decrease and increase in the amount of C16:0 and, C18:1 and C18:2, respectively, as adulterant is increased in concentration.

5.5.2 GC–Mass Spectroscopy

This methodology is similar to GC; however, it is more accurate, reliable, and fast since the two techniques (GC and MS) are integrated to form a single powerful method for analyzing mixtures of chemicals. GC–MS equipment is connected to a computer and uses advanced software that allows for the building of a library of the structures for targeted compounds that are of interest for the particular type of analysis, which can be easily referred to within and among laboratories.

5.5.3 High-Pressure Liquid Chromatography

HPLC is now used routinely for much of the food analysis worldwide. Modern HPLC has many applications including separation, identification, purification, and quantification of various compounds. The major advantage of HPLC is its ability to handle compounds of limited thermal stability or volatility (Macrae 1988). Preparative HPLC refers to the process of isolation and purification of compounds. Important to the processing of samples is the degree of solute purity and the throughput, which is the amount of compound produced per unit time. This differs from analytical HPLC, where the focus is to obtain information about the sample compound. The information that can be obtained includes identification, quantification, and resolution of a compound (Regnier 1983).

Chemical separations can be accomplished using HPLC by utilizing the fact that certain compounds have different migration rates given a particular column and mobile phase. Thus, the chromatographer can separate compounds from each other using HPLC; the extent or degree of separation is mostly determined by the choice of stationary phase and mobile phase. Purification refers to the process of separating or extracting the target compound from other (possibly structurally related) compounds or contaminants. Each compound should have a characteristic peak under certain chromatographic conditions. Depending on what needs to be separated and how closely related the samples are, the chromatographer may choose the conditions, such as the proper mobile phase, to allow adequate separation in order to collect or extract the desired compound as it elutes from the stationary phase. The migration of the compounds and contaminants through the column needs to differ enough so that the pure desired compound can be collected or extracted without incurring any other undesired compound.

5.5.3.1 HPLC in Food Analysis

In order to identify any compound by HPLC, a detector must first be selected. Once the detector is selected and is set to optimal detection settings, a separation assay must be developed. The parameters of the assay should be such that a clean peak of the known sample is observed from the chromatograph. The identifying peak should have a reasonable retention time and should be well separated from extraneous peaks at the detection levels, during which time the assay will be performed. To alter the retention time of a compound, several parameters can be manipulated such as the choice of column, the choice of mobile phase, and the choice of the flow rate. HPLC application to food analysis had been reviewed by Macrae (1988). Folkes and Crane (1988) reported on the application of HPLC for carbohydrate analysis, such as low melting point sugars and oligosaccharides. Procedures are now well established for quantitative determination of carbohydrates in foods, and in many cases have been adopted as standard methods. For complex lipids, those of low volatility, and those whose chemistry is sensitive to elevated temperatures, HPLC is the most useful technique. There is also extensive literature on the use of HPLC for the determination of vitamins in food (e.g., Parrish et al. 1980; Lambert et al. 1985). Other food-related applications for which HPLC seems suited are the resolution of amino acids into their optically active enantiomers and the analysis of peptides from the Edman degradation, in which the amino-terminal residue is labeled and cleaved from the peptide without disrupting the peptide bonds between other amino acid residues.

5.5.4 Microscopic Determination (Microanalysis)

Scanning electron microscopy and transmission electron microscopy laboratory techniques offer a wide range of technologies. These techniques enable expertise to devise innovative procedures for the study of unusual samples.

A scanning electron microscope (SEM) is a microscope that uses electrons rather than light to form an image. There are many advantages to using the SEM instead of a light microscope. The SEM has a large depth of field, which allows a large amount of the sample to be in focus at one time. The SEM also produces images of high resolution, which means that closely spaced features can be examined at a high magnification. Preparation of the samples is relatively easy since most SEMs only require the sample to be conductive. The combination of higher magnification, larger depth of field, greater resolution, and ease of sample observation makes the SEM one of the most heavily used instruments in research areas today. SEM has been used for the determination of non-halal leather in leather products (Mirgani et al. 2009), and it also has the potential to be used for the determination of non-halal food products.

5.5.5 Fourier-Transform Infrared Spectroscopy

Fourier-transform infrared (FTIR) spectroscopy could be used to analyze food samples such as animal fats, chocolate, cake, and biscuits for the presence of non-halal ingredients such as lard. Analyses include characterizing and identifying the differences in FTIR spectra profiles. FTIR spectroscopy with chemometric analysis offers rapid, simple, reliable, and environmentally friendly analytical techniques that can detect and quantify low levels of lard-adulterated food samples (3%–5% detection limit). Spectroscopic methods are an attractive option, fulfilling many analytical requirements such as speed and ease of use. Of those, mid-infrared methods (Wilson and Goodfellow 1994) have recently been applied to the authentication of a range of materials, including fruit purees (Defernez et al. 1995), jam (Defernez and Wilson 1995), olive oil (Yoke Wah et al. 1994), and coffee (Briandet et al. 1996). Che Man and others have successfully used the FTIR spectroscopy in determining quality parameters in edible oils and fats, such as iodine value (1999), free fatty acids (Che Man and Setiowaty 1999a), anisidine value (Che Man and Setiowaty 1999b), peroxide value (Che Man and Mirghani 2000), and aflatoxins in groundnut and groundnut cakes (2001), and in detecting the presence of lard in mixtures with animal fats (2000). Al-Jowder et al. (1997) used mid-infrared spectroscopy for addressing certain authenticity problems with selected fresh meats and reported about semiquantitative analyses of meat mixtures.

5.5.6 Electronic Nose (E-Nose) Technology

The new analytical electronic nose, the E-Nose™, is based on Electronic Sensor Technology; for the first time, there is a vapor analyzer that performs flash chromatography and VaporPrint™ imaging in seconds (Staples 2001). E-Nose provides rapid early identification and quantification of atmospheric changes caused by chemical species to which it has been trained. It can also be used to monitor cleanup processes after a leak or a spill. Studies have shown that E-Nose can be used as a rapid detection of non-halal food contaminants in the food matrix by characterizing simple or complex odors. These instruments could be used for authenticating halal food, non-halal items such as alcohol, and intoxicating materials, and to some extent for detecting whether the slaughtering of animals is following the Islamic slaughtering protocol, which is to be a purposeful act, the intention of which is to take the life of the animal in order to use it as food. That could be, to some extent, detection of blood retention in the meat or determining the amount of iron (Fe) mixed in the flesh.

Furthermore, the potential of E-Nose technology to sense the presence of pathogens in humans can contribute to the early detection of diseases. Recently, medical applications of electronic noses have been explored. The use of a novel electronic nose to diagnose the presence of aflatoxins and other mycotoxins in food has great potential. The relationship of electronic nose analysis and sensory evaluation of vegetable oils during storage was studied by Shen et al. (2001). Capone et al. (2001) has used the electronic nose for monitoring rancidity of milk during aging. Electronic nose technology was also used as a useful tool for monitoring environmental contamination (Baby et al. 2000).

5.5.7 Differential Scanning Calorimetry

Differential scanning calorimetry (DSC) is a thermo-analytical technique for monitoring changes in physical or chemical properties of material by detecting heat changes. Thermogram profiles show the presence of mixed or added substances such as lard in food samples. This technique also provides fast and accurate determinations of lard mixed with other oils or other animal fats.

DSC is an instrument that has been widely used in polymer science for a variety of analyses. The advantages of DSC are that it works rapidly and simply, and much valuable information can be obtained by a single thermogram with a very small sample, yielding accurate results (Wang and Kolbe 1991). Based on the DSC profile, melting point, a point cloud, and iodine value palm oil could be determined quantitatively (Haryati 1999). Marikkar et al. (2001) reported the detection of lard and randomization lard as adulterants in RBD palm oil by DSC. Detection of animal body fat in ghee and butter using DSC has also been reported by Lambelet (1983) and Lambelet et al. (1980), and Coni et al. (1994), respectively.

5.5.8 ELISA Technique

Enzyme-linked immunosorbent assay, also called ELISA, is a biochemical technique used mainly in immunology to detect the presence of an antibody or an antigen in a sample. It has been used as a diagnostic tool in medicine and plant pathology, as well as a quality control check in various industries. ELISA techniques are relatively simple to perform. In halal industries, the ELISA technique has been used for detection of pig derivatives qualitatively in the food samples, such as sausages of various types of meat. These analyses yielded excellent results for detection of pig derivatives in samples.

5.5.9 Molecular Biology Approaches

Molecular biology techniques are commonly employed in research and service laboratories around the world using the polymerase chain reaction

(PCR), which is a technique to amplify a single or a few copies of a piece of DNA, as a primer, across several orders of magnitude, generating thousands to millions of copies of a particular DNA sequence.

Primers (short DNA fragments) containing sequences complementary to the target region along with a DNA polymerase (after which the method is named) are key components to enable selective and repeated amplification. As PCR progresses, the DNA generated is itself used as a template for replication, setting in motion a chain reaction in which the DNA template is exponentially amplified. PCR can be extensively modified to perform a wide array of genetic manipulations. The PCR technique can be used to verify, certify, and monitor most animal proteins and related products for halal authentication efficiently and effectively, as well as identify some other consumer products, such as genetically modified organisms.

Nucleic acids present in food are the characteristic for the various biological components in complex products. Analysis of specific nucleic acids in food allows the determination of the presence or absence of certain constituents in complex products or the identification of specific characteristics of single food components. As DNA is a rather stable molecule, processed food is generally analyzed using the DNA-based method. Due to its specificity and rapidity, PCR is the method of choice for this purpose.

PCR analysis of food includes the following steps:

- Isolation of DNA from the food
- Amplification of the target sequences by PCR
- Separation of the amplification products by agarose gel electrophoresis
- Estimation of their fragment size by comparison with a DNA molecular mass marker after staining with ethidium bromide
- Verification of the PCR results by specific cleavage of the amplification products

A very convenient approach is to perform PCR amplification and verification in one single run by using a target-specific fluorescent-labeled oligonucleotide probe in a real-time PCR system. Real-time PCR requires more expensive laboratory equipment but allows gel-free product detection without the need to open the PCR tubes after amplification. This approach is therefore less time consuming and labor intensive. It implies a lower risk of contamination, and there is no need to use mutagenic staining dyes such as ethidium bromide. Highly accurate quantitative results can be obtained with real-time PCR.

Many procedures of halal authentication using PCR have been developed, such as the method for species identification from pork and lard samples using PCR analysis of a conserved region in the mitochondrial (mt) cytochrome b (cyt b) gene. Genomic DNA of pork and lard is extracted using Qiagen DNeasy® Tissue Kits and subjected to PCR amplification, targeting the mt cyt b gene. The genomic DNA from lard is found to be of good quality and produces clear PCR products on the amplification of the mt cyt b gene of approximately 360 base pairs. To distinguish between species, the amplified PCR products are cut with restriction enzyme BsaJI, resulting in porcine-specific restriction fragment length polymorphism (RFLP). The cyt b PCR-RFLP species identification assay yields excellent results for the identification of pig species.

5.5.10 Chemical Testing

Traditional wet chemical testing has been used in many laboratories to determine food quality. Many chemists rely on wet chemical methods; however, these methods are considered to be non-environmentally friendly as many of these chemicals are hazardous to living organisms as well as the environment.

Testing of packaging materials and microbial testing are also important for any type of raw material or food, and it is especially important for packed food since it can easily spread by local and/or international trading.

5.6 Halal Research and Development in Malaysia

Although the technical aspects of halal research and development, described in the previous sections, are highly technical, they were developed to apply directly to the needs of producers and processors to specifically verify the important societal values embodied in halal food production, certification of food ingredients and additives, as well as for finding alternatives to existing non-halal or questionable (shubhah or mashbooh) ingredients and food processing aids. The role of scientific expertise in food production, especially in terms of halal interpretations, could strongly influence market potentials and business opportunities along the entire halal food value chain, which is extremely widespread and large. A number of institutions such as the Halal Products Research Institute, Universiti Putra Malaysia, the Halal Industry Research Centre, and the International Islamic University Malaysia, along with other Malaysian agencies, are able to consolidate and integrate the opportunities to optimize resources and increase competitiveness to contribute towards the goal to develop Malaysia as an international hub for halal products and services. The application of professional laboratory techniques, based upon sound biochemistry, biology, and chemistry, is a critical scientific link with the improvement of the lives of the Islamic "global community of practice."

6

Achieving Meaningful Stakeholder Dialogs in the American Midwest[1]: Stakeholder Perceptions and Interactions Using Enhanced Place-Based Appreciative Methods

Brenda Groskinsky

Environmental Scientist

Shawn Grindstaff

Facilitator and Process Consultant

CONTENTS

---

[1] The United States (U.S.) Census Bureau defines the American Midwest to include the states of North Dakota, South Dakota, Minnesota, Iowa, Nebraska, Kansas, Missouri, Wisconsin, Michigan, Illinois, Indiana, and Ohio (U.S. Department of Commerce, Economics and Statistics Administration, U.S. Census Bureau 1984).

6.1 Introduction

Since the early 1990s, the U.S. Environmental Protection Agency's (USEPA) Region 7[2] staff have been involved in the science of, and stakeholder participation in, ecosystem conservation.[3,4] For example, the USEPA Region 7's Great Plains Program initiated in the early 1990s was an internal collaborative effort to protect Great Plains ecosystems. Region 7's Great Plains Program became involved in a larger body of federal, state, and local governmental agencies, tribes, and nongovernmental organizations and constituencies who were interested in obtaining a sustainable future for the U.S., Canadian, and Mexican Great Plains, and was titled the Great Plains Partnership (GPP). The GPP started with select members of the Western Governors' Association[5] pondering how to collaboratively protect migratory birds in the Central Flyway Migration Corridor[6] of the North American Great Plains (NAGP).[7] The GPP was an experiment to see if federal state, tribal, and local agencies, as well as nongovernmental organizations and landowners within the NAGP, could work together to create their own sustainable future. The collaborative efforts of the GPP helped to solidify USEPA Region 7's commitment to engage and work with stakeholders, which in turn created lasting relationships that facilitated more useful methods for promoting long-term environmental and human health and well-being (Clark 1996).

A "Vision for the Future" was created by the GPP and provided materials that were used during a GPP Council Meeting held in Omaha, Nebraska, June 21–22, 1996. The vision statement reflected empowerment of local

---

[2] USEPA Region 7 is one of the ten regional offices located in the United States. It is comprised of four Midwestern states: Iowa, Kansas, Missouri, and Nebraska. These four Midwestern states are known as "the Heartland" (Seavey 2003) with agricultural as their main economy and more than 50% of the population being rural.

[3] In 1990, USEPA's Science Advisory Board established ten recommendations to the USEPA for the purpose of enabling the Agency to set priorities and establish strategies for heightened human health and environmental protection. One of the recommendations stated, "EPA should attach as much importance to reducing ecological risk as it does to reducing human health risk" (U.S. Environmental Protection Agency 1990).

[4] The USEPA Science Advisory Board is a federal advisory committee for the USEPA. Additional information on USEPA's federal advisory committees can be found at www.epa. gov/faca (checked September 15, 2018).

[5] The Western Governors' Association is a nonpartisan organization of governors from the Western states of the United States. Additional information can be found at www.westgov. org (checked September 15, 2018).

[6] Additional information on the *Central Flyway Migration Corridor* can be found at www. audubon.org/central-flyway (checked September 15, 2018).

[7] The NAGP extend westward to the foothills of the Rocky Mountains and also include the arid semideserts of southwest United States and parts of northern Mexico. The NAGP contains a portion of the American Midwest but does not include the Midwestern states east of the Mississippi River. There are close to 20 ecoregions and more than ten land cover types across the NAGP including interior grasslands and semiarid desert. Land cover trends of the NAGP have, and will, continue to occur (Ostlie et al. 1997).

citizens. "We the people of the Great Plains Region, share a Vision for the Future...We rely on each other and on healthy ecosystems, prosperous rural communities, and healthy, prosperous cities...we protect the natural and environmental resources upon which all goals ultimately depend" (Great Plains Partnership and Western Governors' Association 1996).

A report entitled "A Way of Life: Great Plains Citizens Talk About Ecosystems" (The Harwood Group, Great Plains Partnership and Western Governors' Association 1996) was commissioned by the GPP and the Western Governors' Association to better understand ecosystems' role in citizens' lives of the NAGP. The report summarized focus group interviews that were held in eight different communities within the NAGP.[8] The report articulated that ecosystems are "a key part of life" for the NAGP citizens (The Harwood Group, Great Plains Partnership and Western Governors' Association 1996). "People who live on the Great Plains take pride in the land that defines the region and what it means to their way of life. They see the land as a key to their prosperity and to maintain their communities" (The Harwood Group, Great Plains Partnership and Western Governors' Association 1996).

William K. Reilly, USEPA Administrator (1989–1992), talked about the Great Plains and pondered the "land debate." He cited, "Our triumphs on pollution controls notwithstanding, we have conspicuously failed to make the same progress on land use...," and also commented, "Land use issues are highly decentralized and inevitably engage the vested and emotional interests of multiple participants." He recommended that in order to "cross the barricades" and "reframe the issues," to "balance public and private interests," requires all of us to "begin with science" (Reilly 1996). The GPP requested the Great Plains International Data Network (GPIDN)[9] to do just that "initiate international cooperation between agencies in the United States, Canada, and Mexico, and to identify environmental, social, and economic challenges facing the Great Plains" (U.S. Environmental Protection Agency Region 7 1999) by starting with science.

In 1996, the GPIDN created, installed, and implemented one of the first Hypertext Markup Language environmental data applications within the USEPA by collecting and providing data, as well as data analysis and information, from and to, a partnership of landowners, nonprofit organizations such as The Nature Conservatory (TNC), government organizations such as the U.S. Geological Survey (USGS), and the USEPA, and academic institutions such as the University of Nebraska. The application was later facilitated by the USGS and provided displays and information exchanges

---

[8] Grand Island, Nebraska; Lisbon, North Dakota; Twin Valley, Minnesota; Great Falls, Montana; Wichita, Kansas; La Junta, Colorado; Pampa, Texas; Eagle Butte, South Dakota (The Harwood Group, Great Plains Partnership, and Western Governors' Association 1996).

[9] The GPIDN was formed in December 1993 at a meeting in Winnipeg, Manitoba, with Canadian and American representatives. Brenda Groskinsky (USEPA) and Wayne Ostlie (The Nature Conservancy) were the U.S. cochairs, and Hartley T. Pokrant (Manitoba Remote Sensing Center) was the Canadian cochair.

of the NAGP data representing ecoregion boundaries, areas of potential natural vegetation, number of rare species by county, occurrences of rare species, and natural communities with respect to landscapes of biological significance, as well as county summaries of air point-source releases from the USEPA Toxic Release Inventory.[10] One of the most significant accomplishments of the GPIDN and the collaborative relationship between the USEPA and TNC was the development of an assessment of biodiversity in the Great Plains and the publication of "The Status of Biodiversity in the Great Plains" (Ostlie et al. 1997), along with the accompanying addendums "Great Plains Landscapes of Biological Significance" (Aldrich et al. 1997), and "Great Plains Vegetation Classification" (Schneider et al. 1997). These accomplishments clearly contributed to "...helping protect the natural heritage of the Great Plains in a manner which is compatible with sustainable economic activities...[and the] cooperative agreement (between USEPA and TNC) has benefited the public both now and in the future," as noted by Ostlie et al. (1997). The research reports were a collaborative feat of both stakeholders and scientists across the entire NAGP. The information continues to be used and referenced more than 20 years after the publication of the three-volume set (Sampson and Knopf 1996; Savage 2011; Scholtz et al. 2017).

6.2 The Emergence of Biofuels in the United States

In 2005, the U.S. Department of Energy (USDOE) released "Biomass as Feedstock for a Bioenergy and Bioproducts Industry: The Technical Feasibility of a Billion-ton Annual Supply" (2005 BTS—U.S. Department of Energy 2005). The document described the mechanisms and potential for biomass use in the production of biofuel as a means to replace other types of fossil fuel sources such as petroleum. USDOE's 2005 BTS was a landmark resource for bioenergy stakeholders and claimed sustainable production of biofuels, creating a vibrant national biofuels industry for the United States. The report cited, "One billion tons of biomass is enough to produce biofuels to meet more than 30% of 2005 U.S. petroleum consumption" (U.S. Department of Energy 2005). Also noteworthy, in 2005, the Energy Policy Act (EPAct) of 2005 (U.S. Congress 2005) was passed, mandating increases in the amount of biofuel that must be mixed with gasoline sold in the United States. The EPAct of 2005 also provided for loan guarantees in the creation and use

---

[10] The USEPA's Toxic Release Inventory is a resource to gain information about toxic chemical releases and pollution prevention activities reported by industrial and federal facilities (U.S. Environmental Protection Agency 1986).

of innovative technologies in the production of renewable energy such as biofuels. As a result, successful production of biofuels became a nationally mandated priority for many U.S. federal agencies. USDOE, USEPA, and the U.S. Department of Agriculture (USDA) were all charged to facilitate major increases in the development, production, and use of biofuels. The mandate required changes in technologies, practices, and land use. Different and potentially significant environmental modifications and impacts resulting from increased biofuel production were inevitable, such as scaled-up agriculture, changes in land management practices, logistics and production modifications, and an increase in USEPA/State and local roles for environmental and human health regulatory and voluntary compliance programs. Understanding and mitigating the potentially controversial impacts was, and is, part of the USEPA's environmental and human health protection mission.

Ironically, 2005 was a year of both critical and somewhat controversial releases of official documents. On the one hand, the release of the 2005 BTS announced the annual agricultural production of a billion tons of biomass for energy was possible (U.S. Department of Energy 2005). On the other hand, the world was "living beyond our means" and damaging significant ecological resources from land-use change calling for "actions to enhance conservation" and the "sustainable use" of ecosystems (Millennium Ecosystem Assessment 2005).

Jump forward more than 10 years to 2016, USDOE announced the release of Volume 1 and Volume 2 of its 2016 Billion-Ton Reports (BT16) entitled "2016 Billion-Ton Report: Advancing Domestic Resources for a Thriving Bioeconomy, Volume 1: Economic Availability of Feedstocks and Maps and Data on the Bioenergy Knowledge Discovery Framework" and "Volume 2: Environmental Availability of Select Scenarios" (U.S. Department of Energy 2016). The BT16 was the third in a series of "Billion Ton" reports developed by the USDOE along with the 2011 Billion-Ton Update (2011 BTS)[11] (U.S. Department of Energy 2011) and the 2005 BT.

The BT16 maintained that by the year 2040, the United States would be able to "produce at least one billion dry tons of biomass resources (composed of agricultural, forestry, waste, and algal materials) on an annual basis without adversely affecting the environment" (U.S. Department of Energy 2016). Additionally, the BT16 noted the report was "not a final answer, but rather a step to help the nation develop strategies for realizing a broader bioeconomy potential," and noted "bioenergy currently is the greatest single source of renewable energy in the United States" (U.S. Department of Energy Office of Energy Efficiency & Renewable Energy 2016).

---

[11] The 2011 BTS is an update to the 2005 BTS and provided "A spatial, county-by-county inventory of primary feedstocks," "Price and available quantities (e.g., supply curves) for individual feedstocks," and "A more rigorous treatment and modeling of resource sustainability" (U.S. Department of Energy 2011).

As an agriculture producing mainstay, the U.S. Midwestern states were expected to experience a significant increase in the production and use of corn-based ethanol and other biofuels, including soy-based biodiesel and bio gasoline (Mehaffey et al. 2012). USEPA Region 7 was aware that many environmental, human health, and regulatory concerns were being voiced, such as the potential for increased pesticide use, increased soil erosion, and the early return of Conservation Reserve Program (CRP) land[12] back into production. Midwesterners also wanted to know if there would be negative impacts of the biofuel production refineries, specifically upon air quality, water quality, and human health.

6.3 The Need for Fossil Fuel Independence and U.S. Energy Policies and Regulations Supporting Sustainable Production and Use of Biofuels

The U.S. provision of emergency aid to Israel during the 1973 Arab-Israel War resulted in an oil embargo retaliation by the Organization of Arab Petroleum Exporting Countries (OAPEC) (Merrill 2007). Following the embargo, as well as other financially charged "political events" (Hammes and Willis 2005), oil prices quadrupled (Merrill 2007). Notably, in 1974, Arthur Burns, the chairman of the U.S. Federal Reserve Bank,[13] pointed out that several factors contributed to oil prices remaining high (even after the embargo was lifted roughly 6 months later) (Merrill 2007). These factors included OAPEC's world oil market influence, high prices of U.S. whole-sale industrial supplies, the short supply of U.S. industrial materials, U.S. oil companies' lack of excess oil production capacity, and finally, the resultant devaluation of the U.S. dollar (Merrill 2007).[14] Additionally, the United States experienced both the oil crisis and severe inflation during this period, creating not only a desire for U.S. fossil fuel independence but also the efficient use of energy. As a result, a whole host of federal policies and legislative initiatives were implemented from the late 1970s to 2006, supporting fossil

---

[12] The CRP is a federal program administered by USDA's Farm Service Agency. Farmers can agree to remove environmentally sensitive land from production in exchange for a 10- to 15-year contract that provides an annual rental payment from the U.S. government. The CRP legislation was signed into law in 1985 and is the largest private, voluntary land conservation program in the United States (U.S. Department of Agriculture 2013).

[13] "The Federal Reserve Bank" is the central banking system for the United States.

[14] Oil prices, in 1970, were quoted as U.S. dollars and "...all currencies were fixed in relation to the U.S. dollar and in relation to gold..." (Hammes and Willis 2005) until August 1971 when "...the system of fixed exchange rates collapsed..." (Hammes and Willis 2005) and flexible exchange rates came into play (Merrill 2007).

fuel independence,[15] thus contributing to the increase of biofuel production in the United States.

In 2001, the National Energy Policy (U.S. Government Publishing Office 2001) was created by the National Energy Policy Development Group,[16] combining energy independence with national security. The National Energy Policy set the stage in the establishment of the Renewable Fuel Standard (RFS) Program, which was created as part of the EPAct of 2005 and added an amendment to the Clean Air Act of 1970 (CAA) (U.S. Environmental Protection Agency 1970).[17] In 2006, President George W. Bush announced his Advanced Energy Initiative (AEI) (The White House National Economic Council 2006) portraying that the United States needed to "address its addiction to oil" (Bush 2006). The RFS was later revised as RFS2 under the Energy Independence and Security Act (EISA) of 2007. The key purpose of the RFS was to specify the total volume of renewable fuel that was to be blended into gasoline based on the volume of gasoline that was sold in the United States each year. For example, the RFS projected that the total volume was to increase to 7.5 billion gallons (28.4 billion L) before the end of 2012 and in 2016, the blending requirement was 36 billion gallons (136.3 L) before 2022.[18]

Under the new legislative requirements, USEPA was/is responsible for issuing and implementing the regulations that mandate the production and use of biofuel, including setting the targets for how much biofuel (corn ethanol, cellulosic ethanol, biomass-based diesel, etc.) should be blended into existing transportation fuels. Additionally, USEPA was/is responsible for developing a lifecycle assessment[19] of greenhouse gas emissions' calculations under the

---

[15] Federal policies supporting fossil fuel independence between 1970 and 2006 included the National Energy Conservation Policy Act of 1987 (U.S. Congress 1987) and the Energy Tax Act of 1978 (U.S. Congress 1978), which encouraged renewable energy generation and tax exemptions for ethanol blends, respectively (Duffield and Collings 2006). Small ethanol plants were offered insured loans from the Energy Security Act of 1980 (U.S. Congress 1980) and ethanol motor fuel excise tax exemptions from the Energy Tax Act of 1978, which also allowed blenders tax benefits (U.S. Congress 1978). The Biomass Research & Development (BR&D) Act of 1999–2000 (U.S. Congress 1999) created an Interagency BR&D Board to facilitate coordination of biomass research between federal agencies such as the USDOE and USDA, and extensions and additional legislations to the EPAct of 2005 from 1992 to 2005, creating 85% ethanol blends (E85) and alternative and flex fuel vehicles, biodiesel use, credits for corporate average fuel efficiency (CAFÉ), and biomass and wind and electricity generation credits (U.S. Congress 2005).

[16] Established by President George W. Bush in 2001 and directed by Vice President Richard Cheney.

[17] The scope and the role of the USEPA for the implementation and oversight of the *RSF* was/is substantial and included added regulatory responsibility such as the calculation and reporting of current and future environmental impacts, as well as an understanding and management of how greenhouse gases reductions were going to be achieved. Goals that were/are set by the *RFS* touch both the energy and agricultural markets.

[18] For additional information: www.epa.gov/renewable-fuel-standard-program/renewable-fuel-standard-rfs2-final-rule (checked September 15, 2018).

[19] As a tool that has "become more prevalent in research, industry and policy" (McManus and Taylor 2015), lifecycle assessment is used by the USEPA to "assess the environmental aspects and potential impacts associated with a product, process or service" (U.S. Environmental Protection Agency 2006).

CAA and finally, the reporting of the current and future potential impacts on the environment from the increased development and use of biofuels to the U.S. Congress every 3 years.[20] In 2007 and 2008, USEPA Region 7 documented the existing environmental laws that were applicable for the construction, modification, and operation of both ethanol and biodiesel facilities in a two volume set (U.S. Environmental Protection Agency Region 7 2007; U.S. Environmental Protection Agency Region 7 2008).

USDOE's National Renewable Energy Laboratory for renewable energy research and development assisted USEPA to visualize the components of biofuels' production, distribution, and utilization that touched USEPA's regulatory role and furthered USEPA's ability to help promote sustainable commercialization with the private sector. The creation of a concept of operations map for a renewable fuel biomass supply chain helped visualize and document the overall system and included the following components: feedstock production, feedstock logistics, biofuels production, biofuels distribution, and biofuels end use (Riley et al. 2007). Each component of the chain contained multiple examples and technology options, and was helpful in articulating the extent of the USEPA regulatory oversight that was required for each of the supply chain's elements.

6.4 Initiation of an Investigation of the Potential Environmental and Human Health Consequences of Increased Biofuels Production and Use in the United States

"Agriculture practices have environmental impacts that affect a wide range of ecosystem services,[21] including water quality, pollination, nutrient cycling, soil retention, carbon sequestration, and biodiversity conservation" (Dale and Polasky 2007). The Millennium Ecosystem Assessment (2005) proclaimed ecologists were already concerned that current agricultural practices in the United States and abroad were creating an environmental disaster before the increase of agricultural production spawned by the increase in biofuels. "More land was converted to cropland in the 30 years after 1950 than in the 150 years between 1700 and 1850, and now approximately one quarter (24%) of the Earth's terrestrial surface has been transformed to cultivated systems" (Millennium Ecosystem Assessment 2005). Dale and Polasky (2007) summarized, "A major concern is that the increased agriculture production

---

[20] As required by *EISA of 2007, Section 204*.
[21] "The services of ecological systems and the natural capital stocks that produce them are critical to the functioning of the Earth's life-support system. They contribute to human welfare, both directly and indirectly, and therefore represent part of the total economic value of the planet" (Costanza et al. 1997).

over the past 50 years has come at the cost of ecological sustainability that will be necessary to maintain productivity in the future." One reason ecological and agriculture sustainability are collectively difficult to attain at the same time is possibly because sustainability of multiple ecosystem services requires coordinated management rather than the management of prolific "independent farm units" (Goldman et al. 2007).

In 1992, 36.5 million acres (14.6 million ha) were enrolled in CRP under 10-year contracts (Osborn et al. 1992). In 2007, "the U.S. had 36.8 million acres [14.9 million ha] (active in CRP)" (Secchi et al. 2008). The long-term projections report by USDA's Office of the Chief Economist anticipated, "Area enrolled in the Conservation Reserve Program (CRP) is assumed to decline through 2009 as high prices encourage the return of some land to production when CRP contracts expire" (U.S. Department of Agriculture 2007). USDA summarized, "CRP acreage is then assumed to gradually rise to its legislated maximum of 39.2 million acres [15.9 million ha] by the end of the projections (2016), with higher CRP rental rates" (U.S. Department of Agriculture 2007).

USEPA recognized they had a role in the development of critical environmental and human health assessment science and needed to work towards understanding how to mitigate the long-term, and potentially unexpected, environmental and human health consequences of the AEI. To assist the USEPA investigate and formalize a strategy to help ensure renewable fuels would be developed in a cost-effective and sustainable manor, the USEPA National Advisory Council for Environmental Policy and Technology's (NACEPT)[22] Energy and Environment Workgroup, as federal advisory committee to the USEPA, was charged to provide its views on how the Agency could best organize and act to encourage the use of renewable fuels and ensure biofuels were being developed in a way that would be sustainable over the long term. The USEPA requested that NACEPT "provide it with a set of recommendations to build the country's energy supply without sacrificing environmental protections and optimize programmatic and regional activities related to national energy priorities focusing on biofuels" (U.S. Environmental Protection Agency National Advisory Council for Environmental Policy and Technology Energy and Environment Workgroup 2007).

In 2007 and 2008, several meetings were sponsored to facilitate USEPA engagement with NACEPT and invited speakers. In August 2007, USEPA Region 7 reported to NACEPT that they had begun a research collaboration with USEPA's Office of Research and Development (ORD) with the intent to initiate a landscape/land-use change research effort focusing on increased biofuels production in the Midwest (Groskinsky 2007).

---

[22] Detailed information about NACEPT can be found at www.epa.gov/faca/nacept (checked September 15, 2018).

6.5 USEPA's Systems-Based[23] Lifecycle Ecological Research to Investigate the Potential Landscape Changes in the Midwest Created by Increased Production of Biofuels

Region 7 recognized that the U.S. Midwestern agricultural states had the potential to be most affected by the increase in biofuels production. Region 7 was interested in identifying optimal mechanisms to both economically and environmentally implement the production and use of diverse biofuels sources. Development of methods that would help identify how to minimize public health issues as well as water and air quality effects, while increasing ecological resilience and land-use change mitigation were critical. Following a collaborative discussion between USEPA Region 7 and USEPA ORD's Ecological Research Program, a consensus was formed nominating the development of a systems-based lifecycle modeling effort using sustainable modeling concepts[24] along with economic and spatial tools. The team was confident that the new modeling strategy would serve as an excellent mechanism in the creation of critically needed predictions of landscape changes, while notably advancing ecosystem services research methodologies (Costanza et al. 1997; Millennium Ecosystem Assessment 2005; Linthurst 2007; National Research Council of The National Academies 2008). The main objective of the research effort was to determine the environmental impacts (or trade-offs) of increased biofuels production in the Midwest through the use of future alternative scenario modeling (Rustigian et al. 2003; Van Sickle et al. 2004; Santelmann et al. 2006). The collaborative research effort was titled "The Future Midwestern Landscapes (FML) Study" (U.S. Environmental Protection Agency Ecological Research Program 2007). The initial step was to identify and select a set of future landscape scenarios reflecting predicted Midwestern biofuel production possibilities.

USEPA recognized the complexity of the modeling effort and therefore "developed a new methodology for constructing hypotheses about the potential effects of future change scenarios on ecosystem services," and then "applied the scoping methodology" as a "proof of concept." The word "scenario" was used "to define a set of driving conditions that will cause change," while the term "landscape" described "the spatially explicit land

---

[23] Dr. George Gray, USEPA ORD Assistant Administrator (2005–2009) stated, "The most challenging, but also the most exciting aspect of focusing on biofuels is recognizing the need to take a systems approach in understanding how this sector can be environmentally and economically sustainable" (Gray 2007).

[24] The collaborative research team was excited to be able to use advanced systems-based, lifecycle modeling methods that would use sustainability concepts going beyond the traditionally used environmental risk regulatory drivers and models. Additional information on USEPA risk assessment tools and databases can be found at www.epa.gov/risk/risk-tools-and-databases (checked September 15, 2018).

cover, land use and land management practices that resulted from a given policy/scenario" (Bruins et al. 2009).

In support of the FML and its development of "future landscapes," several collaborative research investigations were initiated to "identify ecological service implications of feedstock production for the bioenergy industry for a specific set of scenarios" (Cruse 2007). Additionally, unique data analyses and modeling efforts were initiated and used to gain valuable information as input. An energy allocation optimization model (Johnson et al. 2006) was coupled with an enhanced econometric regression modeling effort (Devadoss et al. 1989) to investigate "the dynamics in agricultural and biofuel markets" (Dodder et al. 2011).[25] In addition, the analysis of crop and soil data, in the context of biomass and bioenergy crop and residue production, and their relationship to carbon sequestration, the carbon cycle, and greenhouse gases were investigated and used as data and information supporting the FML (Nelson 2002; Nelson and Schrock 2006; Pendell et al. 2007).[26]

6.6 What Did the U.S. Midwestern Stakeholders Say to the USEPA about the FML?

USEPA Region 7's past experiences with the GPP provided enlightening observations and documentation of dreams for the future. Region 7 anticipated the Midwestern stakeholder observations of increased biofuels production would be extremely useful and provide critical insight for FML. The FML team realized stakeholder involvement and a better understanding of Midwestern values was necessary in order to understand the drivers that were creating the biofuels industry and the agricultural, environmental, and land-use futures. The FML wanted to gain an audience of interested and influential individuals that could provide perspectives from a variety of interests on current Midwestern trends in agriculture and ecosystem services. The FML was interested to know if the stakeholders understood the research objectives. Would it be possible to advance the FML's scientific

---

[25] The modeling premise was to first use energy allocation optimization by characterizing the minimum total cost after knowing maximum energy demands while considering the individual levels of commodities (biofuels, gasoline, diesel, etc.) that were available per sector. Econometric regression modeling would then be used to identify the relationship between the quantity of commodities in comparison with prices. Both modeling environments running concurrently (output is used as input for each model and vice versa) had the potential to converge and thus jointly create an optimal solution linking agriculture and energy.

[26] In 2007, Charles Rice (Kansas State University), during a 3-year appointment with the Intergovernmental Panel on Climate Change (IPCC), was awarded a Nobel Prize for his work, in collaboration with Richard Nelson (Kansas State University), investigating agriculture's contribution, and mitigation opportunities, related to climate change (Kansas State University Center for Sustainable Energy 2007).

mission in ways that would meet the values and needs of the people living in the Midwest?

On November 7, 2007, the FML held a Midwestern stakeholder workshop at Iowa State University's Memorial Union in Ames, Iowa. An introductory presentation was given to the stakeholders describing USEPA's intent to embark on an ecological research effort studying the future of ecosystem services of Midwestern landscapes in the United States, in the context of increased biofuels production. The USEPA described their interest in developing a modeling environment that could predict how ecosystem services would be modified by human behaviors over the next 10–15 years. The stakeholder participants were encouraged to ask specific questions about the research and its potential outcomes. Small groups (five or six people) were established and the participants discussed and answered a set of questions about what they perceived the future would hold. For example, they were asked to describe their views on quality of life and what they value from living in the Midwest. What trends for the future did they find most hopeful or most worrisome? How would they like agriculture to be different? In closing, the FML team gave examples (in graphic map format, with tabular information as well) of what the future scenarios might look like and asked for comments. For example, what would a future landscape of the Midwest look like if the United States was facing a scenario involving an agriculture crisis during a heightened oil dependency?

The list of the stakeholder attendees represented a variety of interests and included the Iowa Soybean Association, the Iowa Biodiesel Board, the Iowa Pork Producers, Iowa's State and the Soil Conservation Committee, Iowa's Natural Heritage Foundation, the Leopold Center for Sustainable Agriculture, the Sierra Club, Ducks Unlimited, USDA Natural Resource Conservation Service, Iowa Department of Natural Resources, Iowa State University, The Land Institute, Kansas Department of Agriculture; Kansas State Research and Extension, Fagen Engineering, U.S. Fish & Wildlife Service Division of Migratory Birds, Missouri Prairie Foundation, and Nebraska's Association of Resource Districts (The Forrester Group 2007).

Summary results of the dialog portrayed the importance of place, as is so important to all of the stories in this book, and is a fundamental value of people across the planet. In their words, "living and raising a family in the American Midwest" was important to many of the stakeholders (The Forrester Group 2007). Living in a rural setting with a diversified landscape was also important. The natural environment is valued. Many of the stakeholders spoke about their quality of life living in the Midwest where they had open space, fresh air, and a sense of quiet and calmness of the landscape. Some of the stakeholders' identified water quality and water quantity as a concern. Several stated that they were hopeful for the potential of renewable energy and improved technologies reigniting and reinvigorating the Midwestern economy. In answer to the question related to the supposition of an agricultural crisis during a heightened oil dependency, most speculated that it would result in a decrease in Midwestern population and furthermore a decrease in congressional representation. USEPA obtained information that was useful for the study but also noticed some of the smaller group discussions had the tendency to stay focused on negative aspects of the current status. Consequently, it became difficult to facilitate the conversation toward a positive discussion. Several stakeholders voiced that they wanted to be recognized and understood for their hard work and way of life. They did not want to be shoved aside. They did not want others to be making decisions for them. And finally, they did not want others (especially the government) to come in and try to change their way of life (The Forrester Group 2007; Grindstaff and Groskinsky 2016).

In conclusion, by using the information that was gained by asking the Midwestern stakeholders to envision their future with a focus on the place where they lived, the FML team was able to identify "the change drivers of concern," which formed the basis of the FML's "future scenarios" development.[27]

6.7 Consequences among Biofuels and Agriculture, Energy, Economics, and the Environment

After significant increases in biofuels production in the Midwest between 2006 and 2009, it was apparent that connections (or linkages) between biofuels and other entities had been created. The use of agricultural food sources as feedstock for biofuels formed a link between biofuels and agriculture. The replacement of biofuels for petroleum as a motor vehicle fuel formed a connection between biofuels and energy. As biofuels became a market entity it influenced both the energy and agricultural markets and thus created a link between biofuels and economics. Finally, as environmental effects from increased biofuels production were beginning to surface, the linkage between biofuels and the environment was initiated. Inevitably, each of the links became intertwined with each other.

Focusing on the early linkages of increased biofuels production with agriculture, energy, and economics, Coyle (2007) provides commentary, noting existing world oil prices and the higher demand for energy is a driver for biofuels production if there is "availability of low-cost feedstocks." He went on to say, "According to the International Monetary fund, world food prices rose 10 percent in 2006 because of corn, wheat, and soybean prices, primarily from demand-side factors including rising biofuel demand" (Coyle 2007).

---

[27] A detailed description of the methodology used to develop scenario modeling of future changes in ecosystem services in the Midwest, with an increase in biofuel production, can be found in the work of Bruins et al. (2009) and a documentation of the use of the methodology in a modeling example can be found in the work of Mehaffey et al. (2012).

Douglas W. Elmendorf, Director of the U.S. Congressional Budget Office (2009–2015), commented that from April 2007 to 2008, "...corn ethanol production along with other factors, exerted upward pressure on corn prices, which rose by more than 50 percent between April 2007 and April 2008... beyond the 1-year period that ended in April 2008, food prices are likely to be higher than they would have been if the United States did not use ethanol as a motor fuel" (U.S. Congressional Budget Office 2009).

As biofuel production in the Midwestern United States became a viable industry for both farmers and producers, a debate surfaced articulating the increase of biofuels production was causing adverse effects on world agriculture and global food availability. The debate became an elevated topic involving farmers and biofuel producers with those outside of the bioenergy markets for years. The controversy was labeled "food vs. fuel" (Tenenbaum 2008).[28] As a rebuttal to the controversy, multiple economic and policy factors could have been influencing commodity food prices including the "declining value of the U.S. dollar, rising energy prices, increasing agricultural costs of production, growing foreign exchange holdings by major food-importing countries, and polices adopted recently by some exporting and importing countries to mitigate their own food price inflation" (Trostle 2008). Zhang et al. (2013) focused on why the range of impacts were so "wide." They specifically articulate that the "reconciliation" of the "systematic differences" in the design of the economic models is needed (Zhang et al. 2013).

Recall that the FML team was interested in capturing and quantifying the dynamics that existed between the agricultural and biofuels markets. Consequently, USEPA and researchers from Iowa State University began an investigation on how the broader energy system drivers affect biofuel production and its ability to be a sustainable resource as an alternative fuel source. Preliminary research results from the USEPA/Iowa State University collaboration summarized, "In order to fully understand the effects of biomass markets, the new and stronger linkages and feedback effects between national- and global-scale energy and commodity markets must be properly understood and identified using an integrated perspective" (Dodder et al. 2011). In 2015, the collaborative effort acknowledged that "development of biofuels in the last decade has been driven by government policies aimed at providing domestic energy security, rural economic growth, implied greenhouse gas benefits and reduced fuel prices" (Dodder et al. 2015). The research results reinforced the concept that there are significant "linkages between agriculture and energy markets," and there are "substantial impacts" on biofuels production when there are changes in crude oil and natural gas prices. Additionally, "the impact of crude oil prices on the demand for biofuels and

---

[28] Controversy arose over the United States', Brazil's, and the European Union's perceived "promotion of biofuels made from food crops" over the need to address existing world food crises during contemporaneous global events and factors such as relatively higher oil prices (Tenenbaum 2008; Coyle 2007).

their feedstocks is much greater than the impact of natural gas prices on the cost of production of corn and biofuels." Additionally, they forecasted, the "lack of cellulosic biomass markets results in an additional push to process more corn ethanol," which "results in an increase system-wide CO_2 trajectory." Finally, their results suggested that "more studies" focusing on the "link between energy prices and agricultural prices" are critical to better understand agricultural markets and their impact on the production of both food and fuel, especially since, predominantly, the biofuels industry has been, and continues to be, policy driven (Dodder et al. 2015).

The number of assumptions and distinct methodologies that are used for lifecycle assessment make it difficult to obtain consistency in the calculations used to determine greenhouse gas benefits. McManus and Taylor (2015) point out that lifecycle assessment is "more prevalent in research, industry and policy...and its use continues to expand as it seeks to encompass impacts as diverse as resource accounting and social well-being." In conclusion, they note, "Carbon policy for bioenergy has driven many of these changes" (McManus and Taylor 2015). "The burden is put on the regulatory agencies to determine the GHG-intensity of various fuels, and those agencies naturally look to science for guidance" (Liska and Perrin 2009). Furthermore, "Even though much progress has been made in determining the direct lifecycle emissions from the production of biofuels, the science underpinning the estimation of the potentially significant emissions from indirect land use change (ILUC) is in its infancy" (Liska and Perrin 2009).

6.8 The Second FML Stakeholder Session

In 2010, USEPA Region 7 and the FML research team wanted to once again hear from their constituencies. The people who lived and worked in the Midwest were directly involved in decisions that were creating land-use change. Region 7 and ORD wanted to better understand the positions of the landowners, the biofuels industry, the state and local officials, the nonprofit organizations, and especially what they valued most from living in the American Midwest. By understanding their values and goals for the future, the Agency knew they would be in a better position to understand what the future might look like, so they could develop a better systematic mechanism for estimating the indirect land-use changes, and the resulting emissions that the USEPA was charged to deliver.

This time USEPA wanted to enhance the stakeholder engagement process. Region 7 learned, from their first FML stakeholder session, the use of classical problem-solving methods by first identifying the problem, had the potential to create dialog that begins with a point of failure, and as a result, the potential to create a negative mood. As an alternative, Region 7 anticipated that by initiating engagement with a focus on the positive, in other words an "appreciative approach" (Cooperrider et al. 2003), the dialog would prove to be more useful, creating an environment that would allow the FML team to gain a better understanding of the following:

1. What stakeholders value
2. Knowing more about how "what they value" influenced their decisions
3. How the place in which they live is important in determining their values

Invitations for the second FML stakeholder session were sent out to a range of professionals who had strong connections to Midwestern land use and included commodity farmers, ranchers, biofuels industry professionals, community leaders, state agriculture resource officials, and environmental group leaders. Region 7 knew that many of the stakeholders were going to be on opposite sides of agriculture, biofuels, and land-use related topics and issues; this was by design.

It was decided that by using an appreciative facilitation methodology, the team would create a listening rather than a presentation and dialog inter-action. The USEPA participants sat along the sides of the room and were not permitted to speak or engage with the stakeholders. In other words, the FML team was not allowed to interject their thoughts and ideas or engage in the stakeholders' conversations. USEPA wanted to see if their silence would change the dialog and create a more useful outcome for everyone involved.

The session was held in a large room at a public library in Kansas City, Missouri, on August 11, 2010. The facilitator explained, "USEPA is here to just listen," and that USEPA was interested in better understanding what the stakeholders wanted the future to look like. USEPA wanted to get a sense of the stakeholders' goals and the decisions that they (USEPA) had made, or were about to make, directly or indirectly, particularly with regard to how those decisions have/would affect land-use decisions, related to bio-fuels production in the Midwest. It was also explained that the information that the stakeholders shared would help the Agency form a more relevant research agenda, and consequently decisions that may follow. The facilitator asked the stakeholders to comment on the following topics:

1. Celebration of stories of the past
2. Proud stories of the present
3. Stories describing their hopes for the future

The stakeholder participants were comprised of landowners and ranchers, a commodity farmer/owner, a cattle producer/feedlot manager, a dairy cow operation manager, an ethanol plant owner, a state regulator, an environ-mentally oriented citizen, an agriculture extension service representative, a feedlot manager, a nonprofit community leader, an agriculture commodity group representative, and two biofuel commodity group representatives.

USEPA witnessed that the celebration stories of the past provided a glimpse into the stakeholders' values while creating a mechanism that bonded the group together. For example, phrases that continued to surface included "face-to-face relationships," a sense of "community," decades of "resilience," "working on the land," mechanisms that they devise and use as a means to "overcome challenges," their "experiences of renewal," self-generated "innovation" providing new ways for success, "leadership," and a long-standing "responsibility to their communities, their future, and the land." One woman described her most memorable story of the past, which was listening to a radio announcement of the first man walking on the moon while she was milking her dairy cows. She described how proud she felt to be living in the United States. Alternatively, several stakeholders were saddened by stories of the past because they realized that the small farm operations, and that way of life of the past generations, were becoming no longer economically viable.

The stakeholders' stories of the present articulated the insertion of technology improvements and better resource management while recognizing that there were a lot of pieces involved in the intricacy of land use. Many of them recognized that urban sprawl was becoming an issue, noting that their new neighbors knew nothing about farming and what it was like to live in open space where one could see the sky and the stars, and not have the constant noise of cars and trucks. They also commented that regulation was a burden that was being inflicted upon them, noting that there were too many regulations and that the regulations were too complicated. One agricultural stakeholder commented that he received a letter of violation for 32 different codes, and, a second letter from the same agency on the same day, that he was a "model of green agriculture." Many of the stakeholders were confused about the "indirect land-use" regulatory requirement for ethanol production and spoke of it as being "unfair." One of the stakeholder participants brought up "climate change" while acknowledging the importance of farming, but then challenged the farming communities to take action at the same time. Stories of the participant's climate change experiences outside of the United States were so convincing that several stakeholder participants agreed to meet at a later time and discuss the topic further; they noted that "EPA" would not be invited to that meeting because they did not feel that they needed "EPA" to develop solutions for them.

Stories of the future tended to focus on trust, education, incentives, and conversation. Several of the stakeholders wanted to create better educational systems for the general public about agriculture. They wanted the incentives that were being offered to be more useful and easier to implement. They expressed that promises made by the federal government, e.g., that they would be able to receive incentives, should be kept. Most importantly, the stakeholders told the USEPA that they wanted more honest conversations on both sides. They wanted to better understand why the regulations were being established, noting that sending a violation notice without a conversation created mistrust. Some of the farmers truly believed that profitability and sustainability could not go hand in hand with each other because there were so many regulations in place. It became clear that many of the stakeholders did not know that USEPA also performed human health and environmental research. Most of the participants believed that USEPA was strictly regulatory. One standout comment that was made that really caught the FML team's attention was "your science is not the same as our science." Several USEPA participants were disappointed they were not able to engage with the stakeholders, and they really wanted to understand the context of that remark in particular.

There was an overwhelming confirmation that the stakeholders had "learned something" from the session, and they specifically acknowledged this. They learned that even though some of the participants were on opposite sides of the topics discussed, honest and positive communication built mutual respect and created a better understanding between the participants. They summarized that follow-up dialog would have the potential to lead to better understanding and integrated planning for all. One stakeholder committed to, and urged others to also, return to their communities as a leader in the creation of dialogs that could help overcome the challenges they were facing. The FML team witnessed many of the stakeholders exchanging contact information for the purpose of continuing the conversation.

The honest dialog that was obtained from the second stakeholder session created insight on how to make the FML research effort more relevant and useful to the constituencies in the Midwest, noting that the FML research agenda was never discussed. The FML team gained better insight on how stakeholders from the Midwest view government and specifically the USEPA. The methods used for the second session provided much more thought-provoking dialog than the first session because of the honest exchange between the stakeholders. Focusing on successes rather than classic problem-solving methods genuinely created dialog that was positive rather than negative. The stakeholders walked away confirming that they were in charge of "creating" and certainly had the "ability to create" a positive future for themselves and their way of life.

"Momentum lost after a forum like this is the biggest missed opportunity for the Agency and most importantly to the Midwest and its future" was a stakeholder comment to the USEPA and FML team during the second USEPA stakeholder session on U.S. Midwestern biofuels production held in Kansas City, Missouri, on August 11, 2010.

6.9 Land-Use Change, Carbon Balance, Greenhouse Gases, and Climate Change

"High corn and soybean prices, prompted largely by demand for biofuel feedstocks, are driving one of the most important land-cover/land-use change…events in recent US history; the accelerated conversion of grassland to cropland in the US Corn Belt" (Wright and Wimberly 2013).

As part of the FML, USEPA developed a methodology to predict what the Midwestern landscapes would look like in 2020 under various biofuel production initiatives and incentives. They predicted, "meeting the needs for biofuel targets under EISA without loss of livestock, animal feedstock, or grain for human consumption would require a substantial increase in production of corn on already-existing farm land" (Mehaffey et al. 2012). "The Midwest, which has the highest overall crop production capacity, is likely to bear the brunt of the biofuel-production-driven changes" (Mehaffey et al. 2012).[29] The study acknowledged that the use of their fine-scale data set (Mehaffey et al. 2011) allowed them to "estimate ecological response functions related to nutrient and pesticide loading and retention rates at a fine scale using accumulation models," and establish a "proactive" mechanism to "maintain sustainable production for both food and fuel into the future" (Mehaffey et al. 2012).

In 2007, during the early stages of increased biofuel production in the Midwest, USDA's acreage prediction for 2016 stated that CRP land was assumed to "gradually rise to its legislated maximum of 39.2 million acres" (U.S. Department of Agriculture 2007). On October 28, 2016, a USDA press release highlighted, "More than Half a Million Americans Involved with Protecting 24 Million Acres" (U.S. Department of Agriculture 2016), which notably was significantly less CRP acreage than what was predicted. "Agricultural conservation policy does not happen in a vacuum but, rather is linked to a myriad of other policies affecting individual farms, national and international trade, and energy production and distribution" (Secchi et al. 2008).

In 2016, USEPA was forced to propose a delay for the mandated 2017 cellulosic volume schedule specified by the RSF because "…real-world constraints in the marketplace needed to supply certain biofuels to consumers, have made the timeline laid out by Congress impossible to achieve" (U.S. Environmental Protection Agency Renewable Fuel Standard Program 2017). Development delay of the cellulosic fuel inevitably causes the mandates to facilitate the use of existing biofuels, e.g., corn-based ethanol, to meet the RFS. Unfortunately, land-use change and greenhouse gas emissions remain to be a concern as well as the continuing fuel vs. fuel debate that specifically targets first generation fuels, such as ethanol (Chen et al. 2016). A potential solution exists in the use of "non-food-based feedstocks" because "cellulosic alcohols offer an

---

[29] Mehaffey et al. (2012) predicted the "greatest increases in corn production" would be in the Midwestern states through the use of continuous planting. In the Midwest, farmers typically use soybeans as a rotational crop for corn (alternating the planting of corn 1 year and soybeans the next). Corn requires fertilized (nitrogen rich) soil to aid growth. Soybeans naturally leave nitrogen in the soil, and as such, farmers do not need to use chemical fertilization for the rotated corn crop. Midwestern farmers might consider implementing continuous planning of corn year after year, without rotation along with the use of chemical fertilizer, if there was a high enough economic incentive to do so.

opportunity to reduce impacts on food supply and price, impose less competition on land use, and further reduce GHG emissions by around 90 percent relative to petroleum-based gasoline..." (Chen et al. 2016).

Cooter et al. (2017) initiated an enhancement to the FML agriculture energy system modeling through the use of the "one-biosphere paradigm" highlighting "production system strengths, flexibility and resilience as well as showing opportunities for further expansion of aspects of biofuels production that form important parts of the bioeconomy." The "multimedia systems modeling approach that explicitly addresses diverging stakeholder interests...combines the economic strength of energy and agricultural markets models with the physical reality of hybrid process-based crop management model to achieve a more complete, system-level picture of biomass feedstock production in the US" (Cooter et al. 2017). The modeling effort added a "more comprehensive biogeochemical accounting...across 20 crops spanning the continental US." A critical research highlight concludes that "technology-driven increases are part of an interconnected biogeochemical system of carbon and nutrient flows that are influenced by management choices." Notably, the model "facilitates the explicit inclusion of the economic and societal factors that influence and, in some cases, control biomass production and food supply outcomes" (Cooter et al. 2017).

"A number of studies have now shown that a biofuel strategy based on corn ethanol and soy biodiesel may indeed be suboptimal in terms of new energy and carbon balances" (Wright and Wimberly 2013). "Once estimates from the literature for process emissions and displacement effects including land-use change are considered, the conclusion is the U.S. biofuel use to date is associated with a net increase rather than a net decrease in CO_2 emissions" (DeCicco et al. 2016). "Agriculture is one of the main contributors of GHGs to the atmosphere, but also has the potential to mitigate global climate change through the adoption of best management practices that reduce GHG emissions and increase C sequestration in soils" (Rogovska et al. 2011). The potential for soil carbon sequestration to mediate greenhouse gas emissions has been known for over a decade (Williams et al. 2004; Pendell et al. 2007).

We summarize and conclude that our ability to influence sustainability of the natural environment necessitates us to work towards gaining a better understanding of the bigger picture of how the individual pieces, i.e., agriculture, energy, economics, the environment, and most importantly, the stakeholders and a true representation of their values, are linked together.

6.10 What the North American Great Plains Stakeholders Said in 1990

The USEPA serves a public mission to protect human health and the environment. We have learned that the achievement of the mission requires input of the constituencies that they serve. Stakeholder perspectives and perceptions are critical for organizational success because stakeholders have a direct affect and influence on the organization's activities, products, and services (AccountAbility 2011). As a result, it must be "assumed that stakeholders are instrumental in achieving organization success," forming the conclusion that "meaningful dialogues with stakeholders are critical" (Grindstaff and Groskinsky 2015).

USEPA's experiences working with stakeholders in the Midwestern United States, while developing the FML, have shown that in order to gain honest and useful dialog, Region 7 and ORD had to embrace a facilitation methodology that incorporated positive approaches as well as an inherent desire to listen. The honest exchange with stakeholders in the Midwest helped the USEPA gain insight on how to create a research effort that was more relevant and useful to the constituencies they serve (Grindstaff and Groskinsky 2016).

In 1990, the stakeholders of the NAGP talked about trust, values, and the reliance on each other. They spoke about valuing the natural environment and their rural lifestyles. They realized that they have freedom and opportunity. They knew change was inevitable, and as a result, they thought about what the future would hold for future generations.

What did the stakeholders from the Midwest say in 2007 and 2010? The vision for their lives included open space and the natural environment, and a responsibility to support their communities, their future, and the land. Even though some of the individuals who were on opposite sides of the topic of biofuels production in the Midwest, they ultimately made commitments to themselves and others to work together by planning future conversations. They recognized that change is always present, and therefore, they value and rely on each other to accept the change, while working toward maintaining their quality of life for the future.

6.11 Going Back to the Beginning with Science and "The Bigger Picture"[30]

"Devising ways to sustain the earth's ability to support diverse life, including a reasonable quality of life for humans, involves making tough decisions under uncertainty, complexity, and substantial biophysical constraints as well as conflicting human values and interests" (Dietz et al. 2009).

---

[30] The phrase, "The Bigger Picture," is quoted from an article posted on The Economist (2009). The article discussed Elinor Ostrom's Nobel Prize for her studies on "human institutions." Dr. Ostrom suggested that methods needed to address critical environmental problems should "include dialogue among interested parties, officials, and scientists; complex, redundant, and layered institutions; a mix of institutional types; and designs that facilitate experimentation, learning and change" (Dietz et al. 2009).

Additionally, William Reilly, USEPA Administrator (1989–1992) stated, "A primary reason for the failure to manage our lands rationally is the continuing lack of consensus about the proper reach of government and the public authority in constraining the behavior of private landowners. Instability and unpredictability have characterized this tension from the very beginnings of European settlement of the North American continent." (Reilly 1996). We maintain that science that focuses on an interdisciplinary approach (or perhaps even the "transdisciplinary approach" outlined in Chapter 1), i.e., one that involves the social sciences, along with the environmental sciences, is more beneficial than just the environmental sciences, alone, primarily because understanding how the natural world around us works, by definition, involves people and their perspectives/knowledge.

In Chapter 7, you will find this book's final example of how similarly energetic people, in concert with environmental science knowledge and professional, have propelled a once-degraded, and indeed a formerly war-torn environment, into an economically viable and environmentally sustainable landscape.

7

Community-Based Management of Mangrove Forests in Southeast Asia

Richard A. MacKenzie
USDA - Forest Service, Pacific Southwest Research Station, Institute of Pacific Islands Forestry

Kristin Jayd
University of Maryland

Hong Tinh Pham
Hanoi University of Natural Resources and Environment

Sahadev Sharma
University of Malaya

CONTENTS

7.1 Introduction

The tropical and subtropical coasts of the world are lined with a highly special-ized type of forested wetland system known as mangroves. Mangrove forests occupy intertidal zones and are adapted to regular inundation by a range of salinities (e.g., freshwater to oceanic) (Tomlinson 1986). The term "mangrove" is a descriptor of function, not phylogenetic relationship; there are nearly 75 spe-cies of mangroves within 20 different families that include small shrubs, palms, and trees (Duke 1992). Several morphological and physiological adaptations allow mangrove trees to survive the harsh conditions of coastal and estuarine life. Their highly vascularized root systems exclude salt from the soil water they utilize and pump oxygen down into anoxic sediments (Tomlinson 1986). Pneumatophores and knee roots project upward from the sediment, whereas prop roots and buttresses extend radially from trunks to provide stability in unconsolidated sediment and areas of high tidal action (Figure 7.1). Roots and leaves are also able to extrude salt to maintain the balance of cellular fluids that allow for normal biological function in saline environments. Mangrove trees also have a specialized form of reproduction through vivipary, i.e., after flower-ing and pollination, fruits germinate into seedlings, called propagules, while still attached to the tree. Propagules can then drop into the water and disperse via oceanic currents to colonize other suitable areas. This combination of traits has resulted in a pantropical distribution of mangrove forests along tropical and subtropical coastlines (FAO 2007; Giri et al. 2011).

7.2 Benefit of Mangroves to Coastal Human Populations

Human coastal populations benefit directly and indirectly from mangrove for-ests. Direct benefits include harvesting mangrove products for food, medicine, fiber, and fuel. Submerged mangrove trunks and roots provide aquatic habitat for resident fish, shrimp, and crabs (Primavera 1998; MacKenzie and Cormier 2012) that are important food sources for people (Naylor and Drew 1998). These resident organisms are also fed upon by larger transient fish that migrate into

FIGURE 7.1
Knee roots and pneumatophores are seen in the foreground rising out of the sediments, whereas prop roots and buttresses can be seen in the background. (Photo: R. MacKenzie.)

mangroves from adjacent seagrass beds or coral reefs at high tide (Vance et al. 1996; Rönnbäck et al. 1999; Nagelkerken et al. 2008). Transient fish are also an important food source for people (Rönnbäck 1999). Above the water, terrestrial species, especially birds, also rely on mangrove forests as habitat (Nagelkerken et al. 2008) and contribute to the resources available to human users of the forest. Honeybees build hives in mangrove forests, capitalizing on their dense floral displays to make honey that is harvested by many coastal communities (Walters et al. 2008). Various parts of mangrove trees are harvested and used as fodder for livestock, such as pigs (Walters et al. 2008), or to prepare a variety of traditional medicines (Bandaranayake 1998; Balick 2009). A number of mangrove tree species provide pest- and rot-resistant hardwood, ideal for the construction of homes, canoes, ships, furniture, and handicrafts. This hardwood is also a valuable source of charcoal or fuel wood that burns hot and for long periods of time (Walters et al. 2008; López-Angarita et al. 2016). Products harvested from mangroves are not only consumed and utilized directly by the people and families that harvest those products but also sold in markets to provide a source of income (Naylor and Drew 1998).

Indirect benefits of intact mangrove forests to human coastal populations include protection from storm events and wave action, increased water quality, and the provision of a natural defense against impacts of climate change, such as rising sea levels. Dahdouh-Guebas et al. (2005) found that areas with intact mangrove forests experienced less wave damage than areas without mangroves, during the 2004 Indian Ocean tsunami. This was due to a reduction in wave height and water velocity as waves moved through dense stands of mangrove tree trunks and aboveground roots. Wave height can be reduced by up to 60% with every 100 m of mangrove forest, though this varies with mangrove tree species, stem and root density, and forest condition (McIvor et al. 2012). Reduction of water velocity by trunks and roots during normal tidal inundation also increases the residence time of water in mangroves and allows plants and bacteria to utilize any excess nutrients in the water column (Alongi 1996). Reduced water velocity also decreases the sediment-carrying capacity of the flooding water, and the sediment is deposited on the forest floor (Furukawa and Wolanski 1996; Furukawa et al. 1997). The lower sediment and nutrient concentrations mean cleaner water that ultimately flows out onto adjacent nearshore systems, healthier seagrass beds, and coral reefs (Golbuu et al. 2003). Mangrove forests can also provide a natural defense against the impacts of climate change. High primary productivity of mangrove trees results in the removal of massive quantities of carbon dioxide, a greenhouse gas, from the atmosphere. More than 90% of this carbon can be stored belowground in roots, where decomposition is minimal due to the waterlogged and anoxic conditions of mangrove soils (Donato et al. 2011; Murdiyarso et al. 2015). Thus, while mangroves represent less than 1% of the global forest cover, they can store up to five times more carbon than any other forested ecosystem (Donato et al. 2011; Alongi 2014). Belowground carbon accumulation coupled with sedimentation that results from reduced water velocities also build up the mangrove forest floor. This allows mangroves to maintain their forest floor elevation relative to sea level and thus their resilience to climate change. This also helps protect coastal communities from the flooding that is expected to occur with sea level rise, though this will vary with forest condition, suspended sediment loads delivered to mangroves, and tidal range (e.g., difference between low and high tides). For example, Lovelock et al. (2015) developed a model that predicts mangrove response to sea level rise until 2100. Mangroves with a tidal range <1 m (3.3 ft) and suspended sediment concentrations <2.5 g/m^3 (0.08 oz/yd^3) could be submerged and lost as early as 2070. Mangroves with tidal ranges of 1–10 m (3.3–32.8 ft) and suspended sediment concentrations greater than 2.5 g/m^3 (0.08 oz/yd^3) were not predicted to be submerged by 2100.

7.3 Exploitation of Mangrove Forests

Coastal human populations have exploited mangroves for the ecosystem goods and services listed above for centuries. The traditional, historical harvest of mangrove forest products was likely small scale and selective, supporting subsistence economies of those coastal communities. Harvesting efforts and impacts would have initially been limited to areas closest to human communities (Walters et al. 2008) and would have posed no threat as the large areas of mangrove could absorb the low-intensity use by humans (Lugo 2002). As populations grew and/or coastal areas were colonized by other nations, demands on mangrove forests grew more intensive. For example, in Latin America, pre-Spanish societies heavily utilized mangroves for timber, charcoal, and fish (Prahl 1989; de Lacerda and Schaeffer-Novelli 1999). When Spanish colonizers arrived, deforestation of mangroves intensified to provide timber needed to build up their naval fleet, resulting in massive deforested areas. After independence from Spain, many countries continued to overexploit their mangrove forests for timber, which resulted in uncontrolled logging activity without replanting (López-Angarita et al. 2016).

Today, demands on mangrove forests continue to intensify as coastal human populations expand and grow, and globalization opens up foreign markets to many developing countries. Large areas of mangrove forests, especially in developing countries, have been cleared for coastal development, aquaculture, and large-scale charcoal production. This has resulted in a rapid rate of mangrove loss over the last four decades that has likely exceeded mangrove deforestation rates of the past (Duke et al. 2007; Giri et al. 2011), though historic rates are difficult to accurately determine. Since the 1980s, Southeast Asia has lost 30%–50% of their mangrove forests (FAO 2007; Giri et al. 2015), while countries such as Ecuador in Latin America have lost nearly 60% of their mangrove forests, mostly to aquaculture (López-Angarita et al. 2016).

7.4 Subsistence and Community Management

Many rural coastal communities continue to thrive today on subsistence-based economies supported by goods harvested from mangrove forests, even more so than communities in urban areas and industrialized nations. These communities are part of the mangrove ecosystems that support them, as their lives are inextricably entwined with them. In these subsistence-based economies, mangroves provide most of life's essentials, and being self-sufficient is often included in cultural identities of mangrove user groups. Harvesting of mangrove products often involves traditional wisdom and practices, familial and ancestral land use and access, and adherence to cultural customs that are handed down through generations (Balick 2009).

The loss of mangroves to agricultural enterprises, such as shrimp farms or oil palm plantations, has negatively impacted subsistence economies and social well-being, resulting in social impacts that have manifested from the loss of traditional lifestyles. In Thailand, which has lost nearly 60% of its mangroves to shrimp farms, social impacts are manifested as a loss of the traditional ways of life. Families previously depended directly on mangroves for food and wood, and indirectly for the health of the fisheries they supported—these important societal foundations that are lost when the mangroves are removed or degraded. When mangroves are removed, families, largely women, are forced to obtain their necessities otherwise. While men can continue to try and fish, reduced or degraded mangroves result in a reduced catch and lower earned incomes. As this has happened, women allocated more hours to working outside the home in places like factories or shrimp ponds. However, their salaries were less than 50% of what men could make from intact mangroves. Women had to work even more hours to try and compensate for wages from deforested mangroves. Less time spent at home by women has significantly reduced their interactions with their families, including lost time with their children, and thus, the amount of time spent sharing and passing down traditional knowledge to the next generation(s) is reduced (Barbier 2006).

The importance of mangrove forests to human communities and subsistence-based economies has been documented in just the past generation (Lugo and Snedaker 1974), and this new awareness has resulted in a shift toward increased mangrove restoration and improved understanding of sustainable mangrove use. Coastal communities often have different ideas of sustainable practices than do managers or policy makers, and concepts of ecological and sociological sustainability often do not match up. The amount of harvested products that are considered sustainable by the harvesters are often not ecologically sustainable, as they would result in a mangrove population decline (López-Hoffman et al. 2006). It is necessary to understand the ecological context of sustainability in order to apply it to the human contexts of economic and social outcomes. Sustainable use of mangroves can be defined as managing the system in such a way that it allows for forest acreages to be maintained or even increased over time, taking into consideration the needs of the current as well as future generations of all mangrove "users" (human and otherwise) of ecosystem services. Harvesting of mangrove products is inevitable as it is an economic necessity (i.e., an imperative mentioned in previous chapters) for rural community members who reside in and around mangrove forests. As such, there is a trade-off between ecological preservation of pristine mangrove forests and meeting the needs of human populations with the resources of mangroves, worldwide, as described in detail in the below examples of Southeast Asia.

Community-based management offers realistic solutions to both sides of the mangrove conservation-use equation. Involving users of mangrove forests in conservation and management plans ensures that their (societal) needs are considered, while improving ecological outcomes, which together can boost the economics of mangrove systems on a range of temporal and spatial scales (i.e., leading to sustainable use). Collaborations between mangrove harvesters/users, foresters, and ecologists support a critical exchange of information and understanding about mangroves, blending academic and quantitative understanding with traditional, lay, and experiential knowledge of the ecosystem. Additionally, fostering intergenerational partnerships between experienced and younger harvesters within the context of long-term monitoring programs could augment the transfer of knowledge between various groups (López-Hoffman et al. 2006). Ecological researchers, governing agencies, and nongovernmental organizations (NGOs) advocate for community-based management in mangrove ecosystems as an avenue for achieving sustainability and the restoration of ecological functions in such a way that includes the needs of the environment, economy, and the people who depend on mangrove forests. Initiatives such as are described in the examples below are becoming more prevalent in the Asia-Pacific region (Datta et al. 2012; Brown et al. 2014), with criteria for conservation success inclusive of measuring site degradation, use degradation, and species-specific degradation. Social sustainability is dependent on appropriate disbursement among members of the community, regardless of their sociocultural status, and this type of management support puts decision-making power in the hands of subsistence-based users and increases its effectiveness (Datta et al. 2012; Rotich et al. 2016).

Community-based management has proved to be a particularly effective restoration method in Southeast Asia. For example, in Indonesia, the Community-Based Ecological Mangrove Rehabilitation (CBEMR) project in the South Sulawesi Province has restored 4,000 ha (9,884 acres) of deforested mangrove forests (Brown et al. 2014). The CBEMR proved to be effective for small and medium extents of mangrove rehabilitation by resolving both biophysical and sociopolitical issues, and underscoring the issues involved with mangrove forest degradation in Indonesia (Brown et al. 2014). Key to the CBEMR success is a community-based restoration project where there is a willingness to participate among all involved, and where there is a thorough recognition of the importance of gender and community rights in mangrove use and planning (Abdullah et al. 2014).

Payment for Ecosystem Services (PES) programs can also work to promote sustainable practices that reflect a synergy of environmental science and the passion of people, creating a landscape inclusive of the environmental, economic, and societal dimensions. For example, governments and NGOs can fund a monetary system of credits based on the capabilities of the ecosystem, in this case mangroves, to sequester carbon. Because of their high rates of productivity, and their close associations with human use and dependence, mangrove forests are good candidates for such programs. PES for mangroves could contribute an additional 2.3%–5.8% to stakeholder incomes, and the community-level impacts of these funds can translate to significant incentives for community participation. However, in some cases, PES could incentivize increased extraction of mangrove products, like investment in more efficient but less sustainable fishing equipment, but these socioecological trade-offs can be mitigated via local, community-level engagement and multilevel governance and participation (Thompson et al. 2017).

7.5 An Example from Việt Nam

The total area of mangroves in Việt Nam decreased from 400,000 ha (161,874 acres) in the 1960s to 73,000 ha (28,328 acres) in the 1990s. This was initially due to defoliating chemicals used in the Việt Nam-American war, forest fire, and collection of fuel wood. However, the rate of mangrove loss was greatest in the 1990s as deforestation significantly increased for the creation of shrimp aquaculture ponds.

Today, the area of mangroves in Việt Nam has increased to more than 200,000 ha (80,937 acres). The recent increase in mangrove area is largely due to restoration efforts and protection policies; nearly 75% of mangrove area has been intentionally planted. Most of Việt Nam's mangroves are located in the southern provinces, with more than 60% located in the Mekong River delta and 20% in the southeast region. Only 20% of Việt Nam's mangroves occur in the coastal north and Red River delta (Hawkins et al. 2010; Que and Dai 2015).

7.5.1 Mangrove Protection and Restoration

Since the end of the Việt Nam-American war in 1975, mangrove restoration has been a priority of the Việt Nam government and local communities. Funding from government and NGOs (e.g., Save the Children UK, World Bank) has supported the planting and rehabilitation of hundreds of thousands of hectares of mangroves along coastal areas (Tuan et al. 2008). The success of many of these restoration projects is in large part due to the participation of communities living in or near mangroves (Figures 7.2 and 7.3). When a mangrove restoration project is approved and implemented, people from local communities are responsible for preparing and caring for mangrove seedlings, identification of planting sites, planting seedlings, and caring for mangroves after planting. Specialized community groups have also been formed that participate in specific conservation/restoration activities (Hong et al. 2008), including the following:

- Mangrove Planting and Protection Group: This group engages in planting and protection of mangroves as well as community outreach to raise awareness about the role of mangrove forests. They also identify suitable planting sites, what species to plant, and developing techniques for planting seedlings based on their own knowledge and experiences in mangrove forests.

FIGURE 7.2
Community preparing mangrove nursery in Nam Phu Commune, Tien Hai District, Thai Binh Province, Việt Nam. (Photo: Nguyen Xuan Tinh.)

FIGURE 7.3
Community involved in mangrove planting in Dong Rui Commune, Tien Yen District, Quang Ninh Province, Việt Nam. (Photo: Nguyen Xuan Tung.)

- Aquatic Collector Group: This group engages in mangrove rehabilitation through such activities as collecting barnacles that cling to mangrove stems that damage mangrove trees. They also inform mangrove protection teams and local authorities about activities that destroy mangroves. They can also directly remind or warn violators about protections in place to prevent deforestation.

- Aquaculture Group: This group plants mangrove seedlings on or around pond embankments to protect aquaculture ponds.
- Cattle and Poultry Farming Group: Water buffaloes, cows, and ducks are often tended to on ponds or on dike embankments, and occasionally in mangrove forests where they can graze. While tending their animals, this group of people collects and provides information on mangrove condition and protection efforts.
- Social Organizations (such as Women's Association, Farmer's Association, and Youth Union): These groups directly take part in decisions about the planting and protection of mangroves. They also raise awareness among members of their organizations and the community at large about the role of mangrove forests and the need for mangrove restoration. They may also deal directly with mangrove protection or restoration by identifying people that have violated mangrove protection policies or by providing funds for mangrove seedling rearing or planting.

7.5.2 Community-Based Management and Monitoring of Mangroves

In Việt Nam, the same laws and regulations are applied to terrestrial and mangrove forest management and monitoring. Specifically, the Ministry of Agriculture and Rural Development, and the Ministry of Natural Resources and Environment are responsible for mangrove management. The People's Committees, representing provincial, district, and commune authorities also have responsibilities for managing local land and mangroves (Hong et al. 2008). However, the management model can be changed to suit different provincial conditions and areas (Que and Dai 2015).

In Northern Việt Nam, the Department of Agriculture and Rural Development (DARD) assigns the management of mangrove forests to the Provincial People's Committee of each commune. Communes can then assign the management of mangrove forests to village communities. Communes and village communities are the smallest management unit under the DARD institutional structure (Figure 7.4). The management and protection of most mangrove forests are not allocated to individual households within communes or communities as they are far from residential areas. Instead, mangrove protection boards are typically set up such that they include representatives from the local government, police, military units, veterans, and village heads who can work part-time. Payment for forest protection can be extracted from the Forest Protection Fund of the Red Cross. Some localities such as Dong Rui Commune in the Tien Yen District and Dai Hop Commune in the Kien Thuy District have piloted models of community-based mangrove management. These management models have initially been effective in managing mangroves in localities.

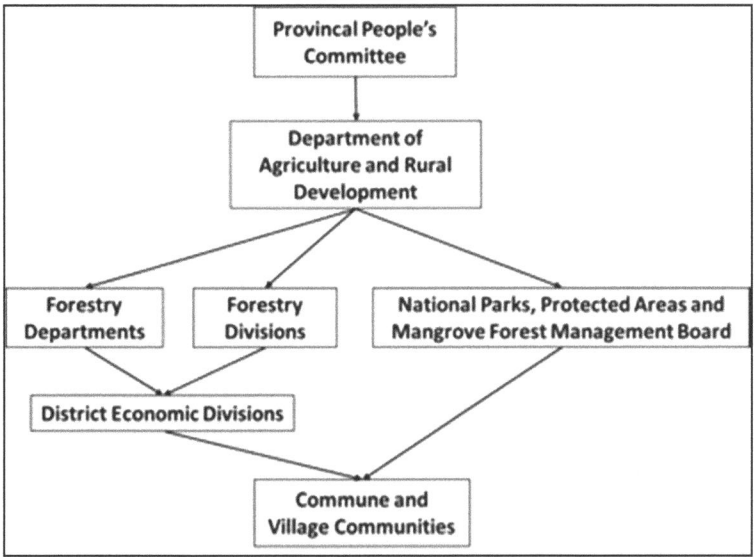

FIGURE 7.4
Institutional structure of mangrove management in northern provinces of Việt Nam (Que and Dai 2015).

In contrast to Northern Việt Nam, mangroves in the Mekong River Delta provinces are under the management of different units, such as Forestry Enterprises or Mangrove Management Boards; communities are not supposed to be involved (Figure 7.5). According to the Civil Code 2015 [(Government of Việt Nam. Civil Code, No. 91/2015/QH13 (24 November 2015)], the community is not a legal entity. This means that the community is not allowed to take part in economic transactions. Instead, the State creates protection forests where harvesting of trees is not allowed. These protection forests are then allocated to the village/hamlet community to protect as stipulated by Land Law, Law on Forest Protection and Development, and other relevant decrees. The State can also generate income to pay local communities for protection. However, communities are not allowed to assign use rights to individuals or community members, and may not be allowed to transfer use, rent, or mortgage rights to community-managed forests. In reality, however, large area of forests, including protection forests, are allocated to local communities for management.

Currently, more than 1% of Việt Nam's mangrove forest area has been allocated to communities for management. In the northern coastal region, most of these mangrove forests are assigned to communities and civil society organizations, such as the Women's Associations or Youth Unions described above. In general, the community enjoys economic benefits from these allocated mangrove forests. Following are some specific models of community involvement in management and monitoring of mangroves in Việt Nam.

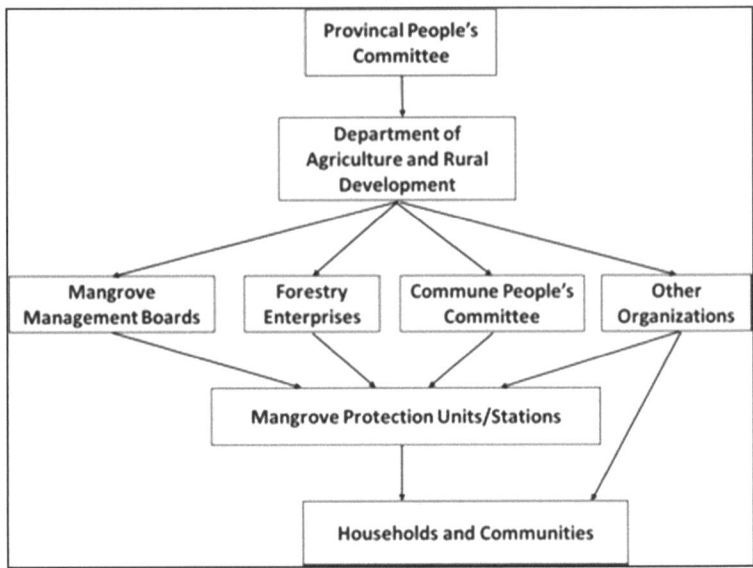

FIGURE 7.5
Institutional structure of mangrove management in Mekong River Delta provinces of
Việt Nam (Que and Dai 2015).

7.5.2.1 Community-Based Mangrove Management Model

This model involves active participation of the community living in or near
mangrove forests. The community-based mangrove monitoring and management model incorporates traditional knowledge, an effective approach in
managing and planting mangroves. This also helps promote the ownership
of mangroves by local people and mobilizes the entire community to plant
and protect mangroves.

The Dong Rui Commune in the Tien Yen District of Quang Ninh Province
is one example of the community-based mangrove management model.
This model was supported from the project entitled "Funding Program for
Small-Grant Projects on Sustainable Management of Tropical Forests in Việt
Nam (EC-UNDP-SGPPFTF)" from 2005 to 2007. The Tien Yen District People's
Committee issued Decision No 368/QD-UB to allocate approximately 1,760 ha
(4,349 acres) of mangrove forest to the local community of Dong Rui Commune
for management and development. Households in the villages of Dong Rui
Commune (Thon Bon, Thon Thuong, Thon Ha, and Thon Trung) then vol-
untarily appointed a Community Forest Management Board (CFMB) within
each village, which would be responsible for the management and protection
of mangrove forests assigned to them. The CFMB and households also agreed

to promulgate a local rule to protect planted and natural mangrove forests. According to this rule, each villager is responsible for

1. Managing and protecting the forest.
2. Rationally harvesting and using it in accordance with the community forest management rule of the village.
3. Supervising the activities of the management boards.

7.5.2.2 Co-Management Model

The co-management model is implemented in state-managed land where the government/authority maintains its management role while assigning ownership, responsibility to protect natural resources, and commitment to sustainable management of land and forest resources to forest owners. In practice, specific rights, objectives, and responsibilities depend on different circumstances and conditions and are often based on the results of negotiations between the authorities and the parties concerned.

"Co-management" is an effective way to maintain and enhance the protective function of mangrove belts, and at the same time provide better living conditions for local communities. In addition, the involvement of local people in mangrove forest management and monitoring helps enhance the resilience of mangroves to the negative impacts of climate change, and accelerate the protection function of ecosystem services provided by mangroves.

The "coastal mangrove co-management" model in the Au Tho B Village in the Vinh Hai Commune of the Soc Trang Province has resulted in the successful coordination of planting, protection, and sustainable sharing of benefits from mangrove forest resources. Prior to the implementation of this model, the management strategy of mangrove forests was to ban access to the forest in all forms as well as harvesting of mangrove trees for fuel wood. The illegal harvesting of wood that ensued resulted in overexploitation of mangrove forests and increased deforestation. Under the co-management model, nearly 300 households, or 40% of the village's population, are allowed to legally enter mangrove forests with special "access cards." These cards allow villagers to harvest firewood in a scientifically proven and sustainable way described in the model's provision on equal sharing of benefits. This model allows people to self-manage and supervise each other for the common goal of protecting mangrove forests for long-term sustainable harvesting, rather than destructive or devastating exploitation, as was the case in the past. This model will help foster environmental protection, as its effectiveness is gaining traction with other villages that are seeing the long-term benefits of this model, and many are becoming interested in balancing the goods harvested from mangroves with the services that these mangroves will continue to provide.

7.5.2.3 *Model of Community-Based Mangrove Management Combined with Livelihood Activities*

This model involves incorporating aquaculture or ecotourism into community-based mangrove management. In the past, farmers used extensive farming and improved-extensive farming systems. Mangrove forests were clear-cut to create ponds that were either naturally stocked or intensively stocked, respectively, with shrimp larvae. These farming methods resulted in the massive loss of mangrove cover that occurred in Việt Nam in the 1990s (Binh et al. 1997; Ha et al. 2013). While the improved-extensive farming system is still the most popular method used today, the government has started to develop shrimp–mangrove integrated farming systems where shrimp ponds are created within or adjacent to mangrove forests (Binh et al. 1997; Hai et al. 2015). In the shrimp–mangrove integrated farming systems, ponds can only represent 30%–50% of the mangrove area depending on the size of the shrimp pond (Baumgartner et al. 2016). Another method used is the intensive farming system where high densities of shrimp are raised in pens built in rivers or waters directly adjacent to mangrove forests and where water circulation and tidal conditions are favorable. As a result, there is no need to clear mangroves (Hai et al. 2015).

An example of combined community-based mangrove management with livelihood activities can be seen in the mangroves of Lien Vi Commune, Quang Yen Town, Quang Ninh Province (Figure 7.6). Shrimp–mangrove integrated farming of greasyback shrimp is considered to be an effective model for mangrove protection and economic efficiency, producing a few hundred tons of shrimp per year and a profit of more than 100 million VND (approximately US$4,000) per year (MERC 2016). The contracted allocation of

FIGURE 7.6
Shrimp farming pond in mangroves in Lien Vi Commune, Quang Yen Town, Quang Ninh Province, Việt Nam. (Photo: Hong Tinh Pham.)

mangroves to aquaculture has prevented illegal and indiscriminate cutting of mangroves. The maintenance of mangrove forests around shrimp ponds also creates favorable shady and oxygen-rich conditions that support the growth of high-value marine species. The mangroves are also an abundant source of food and phytoplankton for cultured species.

7.5.3 Incentives for Community-Based Management

Community participation in mangrove rehabilitation and protection results in economic benefits at varying scales. Community members can receive direct payments from the government or NGOs for contracted protection of intact forests or protection and planting of rehabilitated forests. When funding ends for rehabilitation projects, the restored mangrove forests are evaluated and handed over to local authorities. The community then continues to receive funding to protect and manage these forests (Hong et al. 2008). In addition to payments, communities are allowed to utilize the mangroves that they are charged with protecting. This includes harvesting firewood or collecting fish, shrimp, and snails. Communities can also utilize mangroves for aquaculture, but in accordance with the agreement established with the government. Households and communities participating in mangrove restoration and protection are also given priority in land allocation for aquaculture or are provided with loans to support/perform sustainable livelihoods with minimal impacts on mangroves. The community can therefore increase their income and improve their lives, while maintaining a close tie with mangrove forests.

7.6 An Example from Cambodia

The nation of Cambodia is a part of the greater Mekong sub region, where millions of people depend directly and indirectly on coastal ecosystems. Nearly 75% of Cambodia's mangrove forests are found along the coastal province of Koh Kong, with 16% along Sihanoukville Province, 6% along Kampot Province, and 3% along Kep Province (3%) (FAO 1994). In the early 1970s, Cambodia had 94,600 ha (233,762 acres) of mangroves. By 1985–1987, this area had been reduced to 68,500 ha (169,267 acres). The loss of mangroves was largely due to overharvesting of trees for charcoal production. Since the 1990s, large areas of mangroves have been converted to shrimp ponds, although many of these have since been abandoned (Spalding 2010). Urbanization, resort development, and salt panne creation have also had significant impacts on mangrove forests in Cambodia (Tieng et al. In review).

In 1994, the government of Cambodia banned the cutting of mangrove forest trees by way of a fisheries protection measure and had destroyed many of the charcoal kilns in the region. Despite these efforts, mangrove trees are still illegally harvested and charcoal continues to be produced. Annual rates of mangrove loss exceed the background rate of forest loss (an annual loss of 92,562.7 ha [228,727.4 acres] [0.8%] between 2002 and 2010) and have accelerated from 1.6% between 1990 and 2000 to 1.9% between 2000 and 2010 (Tieng et al. In review). Continued loss of mangroves today has been attributed to the lack of management plans and law enforcement, population growth, and a general lack of awareness. The conversion of mangroves to salt pannes for salt production and shrimp ponds also continues to result in a substantial loss of mangroves (Mastaller 1999; Kosal 2004). In 2015, the area of aquaculture ponds or salt pannes that were once mangroves (14,000 ha, 34,595 acres) represented nearly 20% of the total area of intact mangroves (70,000 ha, 172,974 acres) (Tieng et al. In review).

7.6.1 Mangrove Protection and Restoration

Despite the loss of mangroves in Cambodia, increased efforts have been made to protect existing mangroves. In 1993, the Peam Krasoap Wildlife Sanctuary (PKWS) area in Koh Kong Province was established. PKWS is comprised of a complex of mangrove, coral reef, and seagrass habitats in the water surrounding Koh Kong. The mangrove forests in PKWS are considered some of last intact and most pristine forests in Cambodia that still support high levels of biodiversity and continue to provide many ecological functions (Bann 1997; Torell et al. 2004). Within PKWS, Koh Kapic was also established as a Ramsar Wetland of International Importance. Preah Sihanouk National Park and Ream National Park were also created in Sihanoukville in 1993, and include mangrove forests.

The number of mangrove rehabilitation projects has also recently increased in Cambodia. Most of these efforts have been funded by international NGOs. As in the Việt Nam example above, success of such projects in Cambodia is largely due to active participation by the communities that live in or near mangrove forests. These communities are paid to help restore and then protect rehabilitated mangrove forests. As a result, they benefit through direct payments for their work, as well as the ecological goods and services (i.e., ecosystem services) that rehabilitated mangroves continue to provide to them.

7.6.2 Community-Based Management and Monitoring of Mangroves

The success of mangrove protection and rehabilitation hinges on the support and involvement of local communities. The inclusion of these stakeholders improves the position of impoverished rural communities who have been denied the fundamental rights to participate in decisions that impact their well-being and livelihoods. In Cambodia, this approach has been met with some success in reducing deforestation (Spalding 2010). For example, village-level committees formed to manage forestry and fisheries have helped prevent the loss of mangrove forests in Koh Kong Province and around Ream Krasop Park. Mangrove forests continue to be threatened from deforestation as the need for food and money for day-to-day survival often exceeds the perceived benefits or need for conservation of mangrove forests (Nasuchon and Charles 2010).

Many international NGOs have played an important role in community management and monitoring of mangrove forests in Cambodia and are described briefly in the following sections. The projects that these organizations support have also included studies to understand how mangrove protection and rehabilitation can improve various livelihood and socioeconomic issues of coastal communities' activities.

7.6.2.1 The International and Socioeconomic Issues of Coastal Communities' Activities

The International Development Research Centre (IDRC), which funds research in developing countries to promote growth and reduce poverty, worked with the government of Cambodia's Ministry of Environment to initiate Participatory Management of Mangrove Resources (PMMR) in PKWS, Koh Kong. This area was once an isolated and relatively unpopulated province. Until the early 2000s, Koh Kong could only be accessed by boat. After the fall of the Khmer Rouge in the late 1970s, many families migrated into Koh Kong to extract resources (Bann 1997; Marschke 1999). Most villagers engaged in subsistence activities and practiced some form of fishing, while a limited number of wealthier individuals were involved in illegal logging, charcoal production, or shrimp farming. As the population continued to grow, mangrove cutting increased for charcoal production and shrimp farming. In the 1990s, increased trade with Thailand resulted in even more intensive illegal charcoal production and use of illegal fishing gear. Soon, communities noticed that there were less fish and mangroves (Bann 1997; Marschke and Nong 2003) (Figure 7.7), and local fishers were also unable to compete with illegal fishers that used more advanced and sophisticated fishing equipment. This resulted in increased conflict between artisanal fishers and larger boats. While villagers felt there was little to be done to stop Thai or Vietnamese outsiders from illegal fishing, they felt that they could play an important role in other local resource management, such as mangrove forests.

PMMR was initiated in 1997 in two villages with the aim of halting mangrove deforestation, restoring mangrove forests, and improving the livelihoods of coastal communities (Marschke and Nong 2003). This was accomplished by first providing mechanisms to link communities with different levels of government to plan and manage their mangrove forests. PMMR also provided opportunities for communities to express their concerns and needs so that livelihood options could be considered. This model was initially difficult to implement as authoritarian leadership was the model often used in Cambodia. Over time, however, communities became more

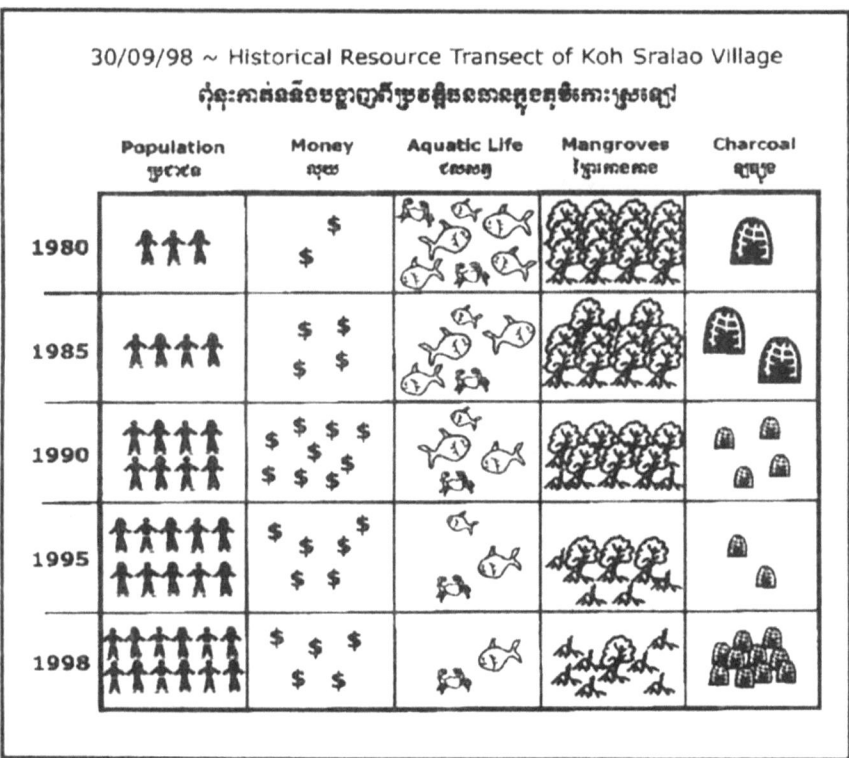

FIGURE 7.7
Change in resources in Koh Sraloa Village. Communities noticed that as populations grew, so did the use of mangroves. While increased money initially came in, this was a short-term gain that resulted in fewer fish and mangroves. (Marschke and Nong 2003.)

comfortable with participatory management and working committees were formed, comprised of interested individuals, members from each section of the village, and others with key knowledge. The committees regularly met to identify management needs, design management regulations, and draft management plans. Action plans were then drafted with staff from IDRC, with less controversial plans often receiving support from other villagers. Action plans that directly impacted villagers were more challenging to implement and required strong leadership from village committees as well as an adaptive approach to revisit plans until they could meet the needs of villagers. A lack of a legal framework to support community-based management in Cambodia further complicated implementing action plans, though there was informal support from the Ministry of the Environment. A lack of a legal framework resulted in complications for getting support at multiple government levels as well. For example, local officials or governors interested in supporting a community-based action were reluctant to support that action without a clear legal mandate or policy.

One example of a village's action plan was one that proposed replanting mangroves. Many villagers participated in the action plan, which resulted in the replanting of more than 10 ha (25 acres) of mangroves (Figure 7.8). However, only a few families (about a dozen out of 300) were involved in patrolling the mangroves because of the risk of violence. Other action plans included active mangrove patrolling, environmental education activities, and home gardening. Many of the community members that were once involved in illegal activities such as charcoal production could now be engaged to patrol and protect mangrove forests. Funding was also provided to build outposts to monitor illegal harvesting of fish and illegal charcoal production.

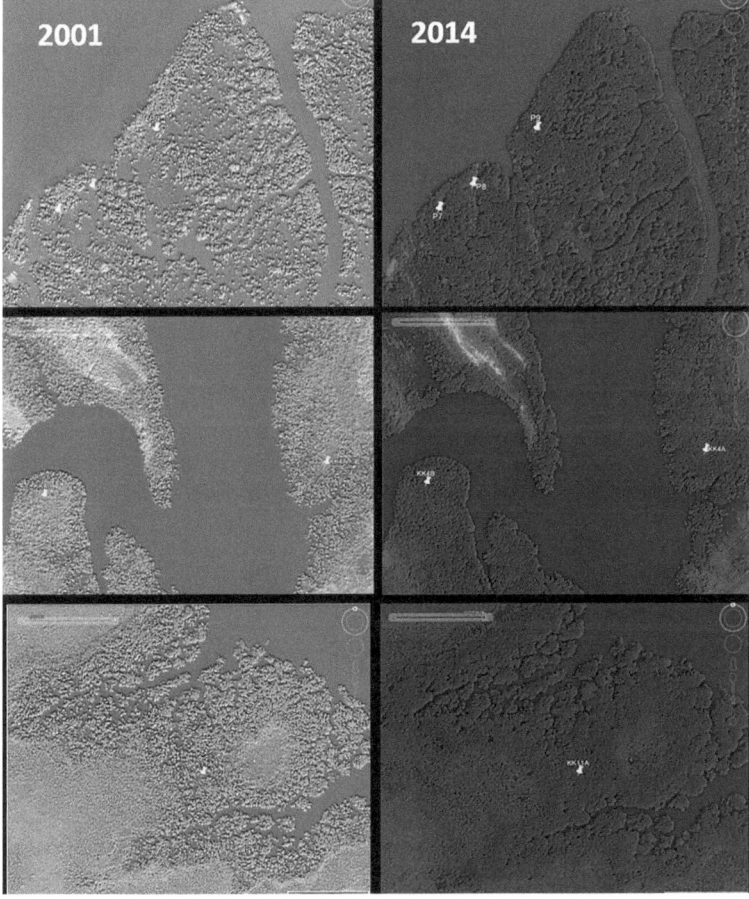

FIGURE 7.8
(See color insert) Images showing a successful mangrove outplanting that occurred in 1997. Yellow pins are in areas that were replanted, and mangrove canopy cover can be seen increasing from 2001 to 2014.

It is hoped that other communities interested in mangrove management and sustainable livelihood issues can learn from the successes of others, and further, this initiative outside of Koh Kong; PMMR will continue to work with other coastal communities in managing and protecting mangrove resources. The work conducted in PKWS has already increased the awareness about the importance of mangrove forests and the need for protection now and for future generations. This collaborative approach has promoted community participation in natural resource management in other areas, such as the community fisheries project sponsored by FAO in Siem Reap and Ream National Park in Sihanoukville (Marschke and Nong 2003).

7.6.2.2 Australian People for Health, Education, and Development Abroad

Kampot Province lost approximately 50% of its mangrove forest in 1992–1993 from firewood exploitation and the creation of salt pannes and shrimp ponds. This boom in mangrove exploitation was attributed to local poor people lacking income after the long civil war with the Khmer Rouge, increased demand for firewood and charcoal from neighboring countries, and a fear of harvesting wood in upland areas due to land mines (Paul 1998).

In 1995, Australian People for Health, Education, and Development Abroad (APHEDA) started a project to reforest mangroves along the Kampot Province coastline (M.R.C. 2018). Working with the Provincial Department of Agriculture, APHEDA established a Community Mangrove Forestry (CMF) group that included 26 families from the Kampot District that planted locally available *Rhizophora mucronata*, *Rhizophora apiculata*, and *Xylocarpus granatum* seedlings. Seedlings were planted on Koh Smao Island, and after 30 months survival rates were 70%. The group also implemented awareness programs and projects to educate local coastal communities on the importance of protecting mangrove forests from cutting, and they worked to motivate community members to restore and protect mangroves. Other outcomes of community meetings were plantation programs in 26 villages, with the goal of planting 60,000 *Acacia auriculiformis* seedlings, as an alternative fuel wood. Interviews with some of these villages suggest that mangrove deforestation that resulted from firewood harvest was reduced by 80% because of these alternative fuel sources (Paul 1998).

7.6.2.3 Culture and Environment Preservation Association

Culture and Environment Preservation Association (CEPA) is an NGO that was established in 1995 by the Royal University of Phnom Penh. In 1997, CEPA worked with villagers from Koh Kong to restore mangroves in PKWS. With funds from Canada, CEPA and the community were able to plant 1 ha of mangroves. Funding has since stopped, and CEPA is no longer involved in the project. However, yearly monitoring of the 1 ha (2.5 acres) of mangroves has continued to occur (M.R.C 2018).

7.6.2.4 Mangroves for the Future

Mangroves for the Future (MFF) is a regional initiative program that promotes conservation and sustainable development of coastal ecosystems. MFF uses a partnership-based, people-focused, policy-relevant, and investment-orientated approach to promote healthy coastal ecosystems. This approach also builds and applies knowledge, empowers communities and other stakeholders, enhances governance, secures livelihoods, and increases resilience to natural hazards and climate change.

In 2012, Cambodia became an MFF outreach country. Since 2014, MFF has initiated mangrove conservation, management, and restoration in the Koh Kong Province through an MFF—Small Grant Facility (MFF-SGF). The main objective of MFF-SGF is to improve local communities' natural resources and thus their livelihoods. MFF-SGF worked with the Department of Environment (DoE), PKWS staff, and local authorities to plant mangroves in the Toul Korki mangrove-protected area. During this program, 25,000 seedlings were planted across 4 ha (10 acres), with the participation of 67 local community members.

MFF-Cambodia has also started a farm-based incentive for local coastal communities in Cambodia. The Sustainable Livelihood through Improving Ecosystem in Mangrove Area (SLIEMA) project was initiated by MFF-SGF to reduce overfishing in PKWS by involving local communities in identifying sustainable ways to collect and farm fish. Research and Human Resource Development also provided multiple trainings to provide solutions for farming challenges that local communities were facing.

7.6.3 Incentives for Community-Based Management

Cambodia's rural communities can be engaged in managing mangroves with the appropriate incentives (Vathana and Penh 2003). In addition to the funds that are paid to communities for planting and protecting mangroves by various NGOs, communities have also benefited from improved livelihoods and cultural identity. Community patrols have reduced illegal harvesting of charcoal and thus mangrove deforestation; increased forest cover that has resulted from these efforts and from mangrove planting; and has increased the numbers of fish, shrimp, and crabs that the communities now sustainably harvest for food (Marschke and Nong 2003).

Another incentive provided to communities to minimize illegal harvesting of mangroves for charcoal was the distribution of alternative fuel stoves. MFF provided manure-powered biogas reactors to 12 local farms in Koh Kong Provinces. These biogas reactors provide energy to local communities, reducing the need to make charcoal from mangrove wood. Manure is also a local and sustainable by-product from farms that saves US$2.50 a day in fuel costs (I.U.C.N. 2016).

Finally, efforts are under way to promote ecotourism, especially in the Koh Kong Province (Reimer and Walter 2013). Funds generated from tourists that have come to the region to see PKWS or other parts of the mangrove forests support communities as well as continued conservation and management of the coastal resources at the grassroots level. These funds are also used as a source of income for mangrove restoration and conservation (Government of Cambodia 2018).

7.7 Conclusion

Mangrove forests across the globe provide vital ecosystem services for humans and nature, especially in the Asia-Pacific Region. The Asia-Pacific region is home to several developing countries, where environmental, economic, and social sustainability are threatened by rapid population growth and climate change. Mangrove forests are experiencing significant challenges related to conservation and management, which are likely to be compounded by future climate change effects. Community-based management can be an effective approach to the conservation, restoration, and sustainable management of mangrove forests. Several successful examples of community-based management from Việt Nam and Cambodia were presented in detail in this chapter, and their effectiveness is continuing to be achieved through inclusion and involvement of the communities that once overexploited them, and have now become their custodians. Participation from these user groups is vital; however, long-term success and sustainability are still in question. Do restored mangrove forests that result from community projects provide the same quantity and quality of goods and services as the intact mangrove forests that they once were? Time will tell us the answer to this important sustainability question. Restored mangrove forests can reach similar levels of biomass, carbon stock, and carbon accumulation rates as the intact mangroves once provided. Among the several critical questions that science is trying to answer about mangroves, in order to guide current and future restoration efforts, is "do restored mangrove forests provide similar fish habitat or storm protection as intact mangrove forests?" Answers to this set of important questions will indeed help us all to better understand how to build sustainable socioecological systems, which serve the needs of communities.

Long-term monitoring, community-focused research, field team training, and project leadership can help to address these questions, and further bolster the success of community-based management so that we can continue to put the power of conservation into the hands of stakeholders. This information can also guide future management efforts and equip coastal communities, the custodians of these magnificent forests, with the insight and acumen to sustain them into the future.

Disclaimer

The research, analysis, and other work documented in this chapter was fully or partially funded by the USDA Forest Service; however the findings, conclusions, and views expressed are those of the authors and do not necessarily represent the views of the USDA Forest Service.

8

Common Themes and Lessons from Our Global Stories of Collaboration and Transformation

Ricardo D. Lopez

USDA - Forest Service, Pacific Southwest Research Station

CONTENTS

After spending time exploring the last seven chapters, and journeying around the world, there are likely several communities and/or environmental science topics that resonated with you. This chapter is meant to simply reflect a bit on some of the lessons that we have learned from the chapters in this book and some of the common themes that are threaded throughout all of these stories of collaboration and transformation. The power of collaboration is evident among all of the lives and topics covered, leading us to a new set of perspectives on not only the transformation of environmental conditions around our changing world but also the transformation of relationships that is currently happening as a result of collaborative approaches to the environmental sciences. As one ponders more deeply the work associated with the diverse collaborative approaches taken among the geographies within this book, it may make some sense to think of the environmental work you do as not only occurring on a physical landscape but also occurring across a "human landscape," which is very complex and multidimensional. As a whole, this complex integrated multidimensional human and physical landscape can be overwhelming to operate within and, consequently, very difficult to effectively and efficiently accomplish goals within. This is understandable, in that there are so many (biophysical and sociological) factors that must be assessed, managed, and coordinated simultaneously (i.e., scientific, management, political, cultural, and other issues) in an integrated human and physical landscape. However, if you think of the several examples provided in the previous chapters at a fairly high level, a number of themes can be extracted and utilized to effectively guide you through this complex multidimensional landscape of environmental information, and human relationships. As you proceed in this chapter, please take some time to reflect back on the previous chapters, and the challenges the participants overcame in this regard, which may assist you and your team in moving toward more successful collaborative approaches, thereby transforming your project's dynamics, toward a more effective way to address the needs of all project partners.

8.1 Be "Of Service": The Confectioner's Communication Model

Initially, in this chapter, it may be beneficial to say a few words about being "of service," the value of true collaboration when on a team of scientists and nonscientists, and the value of meeting the needs of others, such as partners or clients, with regard to any environmental topic. It should now be apparent after reviewing the many stories of global collaboration and transformation in this book that the communication methods used by the parties involved in an environmental project are one of the key factors in a successful outcome. As demonstrated throughout the global stories of collaboration and transformation in the previous chapters, the somewhat subtle elements of communication and collaboration in a science project become evident when we focus on them and highlight their role in the potential or actual success of the work. To this end, the previous chapters were designed to de-emphasize the science and technology, although each story had a very in-depth science element, opting for heavier emphasis on the communication of participants. Consequently, the stories in this book provide much more emphasis on the human dimensions of the work, as would be expected considering the primary focus (and title) of this book.

The scientific and technological accomplishments of the work in the previous chapters is professional and well-organized science, yet the scientific aspects of the work were predominantly and intentionally "put in the back seat" so that the human aspects of the work could be highlighted more intentionally, and discussed more readily than is normally the case in most environmental science books. Although scientists and practitioners do not always focus on (or explicitly articulate) the societal dimensions of environmental work as intently as this book does, it is very intuitive to some that the effectiveness of communication that one exercises in their work is directly correspondent with the success of how well that scientific work is applied, and indeed accepted by people and communities for which the work is intended to help.

A simple and understandable example that explains why the more humanistic approaches to environmental science are most effective is the analogy of a store, let us say a candy store, and the storekeeper, the confectioner, that operates such a candy store. This analogy applies to collaborative work, and to the human interactions and the actual words and actions that we use during such collaborations, when working on environmental issues. This "candy store analogy" is a simple vehicle that can be useful for understanding the simple parameters of what is necessary to be "of service" when working with partners and communities.

A candy store typically has a prominent sign out front proclaiming that it is open for business and the patrons walking by who (1) desire some kind of candy walk directly in, intent on their needs, immediately searching the bins for what candy they need. There are also people who (2) walk in to a candy store for other reasons, such as simple curiosity; they do not typically eat or buy candy, but they are just in the neighborhood and need a bit of a respite from their stroll, or diversion, so they meander in to look around at what sweets you have for sale—they have no concerns about hanging out in a candy store, but they may simply be interested in solely browsing over the candy bins without purchasing anything. And finally, there are those customers who (3) are lost and did not read the sign at all, and do not really care about candy, but they come into the store anyway and seem a little out of place, but they stay a short while and appreciate the air conditioning.

As environmental scientists and those who work with them, we know there are candy store parallels for the circumstances described above, for our partnership interactions, analogously and respectively as: (1) those people who are wondering if their needs can be met by science, but they do not really know much about how, so they investigate by reading about, or asking a science group, which has laid claim to some specific type of scientific expertise that interests them, without really knowing what specific types of information they really need to know; (2) those partners that are really not looking for a science-based answer, but do not mind listening to new ideas, and yet they have not real expectations from the interactions with scientists; and (3) those "partners" who are truly not interested in scientific approaches, but they listen anyway and tend to not come back after their initial encounter.

The candy store analogy at this point may seem a little trivial or odd, and a bit of a caricature of all users of scientific information. Allow me to more fully illustrate the analogy, tying into this analogy the manner in which we as scientists operate when confronted with the above potential partners, i.e., whether or not we react by being of service, or not. The scientist, analogously, is similar to the "confectioner," because they have a product or perspective to offer the public, and they have finely tuned that product, and are rightfully proud of this fact. It takes a lot of work to produce scientific data and analyses. However, not all candy stores succeed, for a number of reasons, and indeed not all scientific information (no matter how compelling the information may seem) is of interest to the public. By way of this continued analogy, I will provide several very important reasons why candy stores do not succeed, and analogously, why scientist–community partnerships do not develop or flourish to their full potential.

One reason candy stores do not fare well is that the candy they sell is old and thus bad tasting, and really nobody wants old bad tasting candy. Another reason why candy stores do not succeed is that, although the candy is fresh and tasty, it is not the kind of candy that customers need (e.g., selling a lot of hard candy when customers need chocolate). Another reason why candy stores do not succeed is that, although the candy is of the type that the customers need, the store cannot deliver enough to meet customer's demand, or the store takes too long to provide the product, and the customer goes to some other store that is more efficient at restocking the shelves. Another reason why candy stores do not succeed is that the wrapper of the candy indicates a flavor that is liked by a customer but, when the customer opens the candy and tastes it, it is not the flavor advertised and they vow to never return to the store. Customers may subsequently complain to their friends, ensuring that others will not make the same mistake of going to the candy store either.

As scientists, and those that work with scientists, we know these situations fairly well, analogously and respectively, as those where (1) scientific information is outdated or outmoded in such a way as to make it difficult to link to current and relevant issues within communities; (2) science information that is perfectly useful and applicable to community needs, but the information comes too late to practicably meet their needs, such as for a particular decision deadline or event; and finally, (3) the situation where scientists make plans to work with a community, and they agree to provide solutions to particular needs, but ultimately they provide information that is (although very thoroughly studied and articulated) about the wrong topic or not specific enough to answer the questions originally posed by the community.

Whether scientist or nonscientist, we all have probably found ourselves in one or several of the situations described directly above, and have been frustrated by this experience, no matter what your role was in the process. I have met many scientists in my career who have expressed their frustration about their perception that "I did all of that work on the project and nobody in the community is using the information," and community members that have expressed their frustration that "the scientist did a lot of work, and it is all good stuff, but it isn't what we asked for." Again, these general statements are very broad, and I do not intend to cover the full gamut of intricacies and breadth of the many times these mismatches occur between "scientist" and "information user;" I am merely describing these general situations because they describe the foundation of why many partnerships in the environmental sciences fail. All parties are typically genuinely trying to make the science–community partnerships work; yet, they often fail. The main issue is that, because there is a fundamental gap between the expectations of the partners and scientists involved, both scientists and nonscientists are often disappointed in the outcomes.

Luckily, there are numerous examples in the previous chapters on how this potential miscommunication was averted or overcome. The work in Chapters 1–7 all describes great challenges in terms of the needs of the community being served, and the environmental scientist attempting to address the needs, and their techniques for communication and cooperation, and ultimately the successful application of science to those needs. Indeed, all of the examples described in this book exhibit the critical elements that solve so many of the problems that arise within a partnership, which are (admittedly) somewhat caricatured in the candy store analogy, pointing us to the two simple (but often overlooked) maxims of successful project teams:

A. Emphasize <u>effective</u> communication
B. Be dedicated to using a truly collaborative approach to problem-solving (i.e., cooperation)

It should be also noted that the works outlined within this book, although meticulously summarized, are just summaries of a tremendous number and scope of human interactions that occurred over years. The sheer amount of energy and dedication to develop the partnerships described, solving real-world problems like we have described in this book, should be pondered, as we conclude this compilation of the work of the people in all of the localities described (i.e., the many thousands of people who collaborated to accomplish so much, for the sake of their societies). Consider the difficulties that the people within the described communities faced, and note that these are but a small handful of the vast number of environmental projects that are currently under way around the world. Each of the circumstances grappled with in this compilation of stories initially had great potential for miscommunication and failure, given the information needs of everyday people and the prospective work of the scientists involved in the project, but they all arrived at a solution of some sort, albeit not always perfect, but a solution nonetheless, which met the needs of the people within the communities described. This is all to say that partnerships take time, and relentless dedication is required for them to flourish and persist.

8.2 Emphasize Trust and Technical Capability

Among the global stories of collaboration and transformation in this book, you may have noticed the specific (and often very intentionally focused) communication and interaction techniques used, as exhibited by how people organized themselves, interacted with each other, and expressed themselves. The fact that they did interact is key, and from my own experiences, such project meetings and interactions are not always blissful love fests, where immediate progress is made on the topic at hand. Indeed, in

such initial meetings, there can be what seems to be preliminary negotiations and ongoing needs for facilitation, and oftentimes, these interactions
lead to difficult discussions and "hard truths" about issues that are not
seemingly related to science by some, but rather more about how people
feel, prior to the commencement of any work. Applied science, after all, is
where the rubber meets the road, and for the rubber to meet the road it often
requires that the parties involved negotiate and talk about details, such as
the practicalities of timelines, money, outcomes, and the need for relevancy
of the information being provided. These discussions can be difficult, and
often can lead to dealing with issues of trust between or among parties,
or the lack of trust that may exist among the collaborators. For example, in
Hawai'i, during the day-to-day interactions among the Rapid 'Ōhi'a Death
(ROD) Working Group members (Chapter 1), there were times when "check-
ins" were needed for a number of the meeting members if group members,
or I as the lead for the group, felt that some participant's interests may not
have been entirely met during formal meetings or conversations (e.g., intu-
ited from facial expressions or other nonverbal or verbal communication of
the potentially upset party(ies)). For all of the grace and positive work of the
ROD Working Group, there were some instances of not-so-nice behaviors,
which required attention by those who were leaders (either formal or informal) of the working group or subgroups. An example of this type of check-in
was when the Merrie Monarch Festival activities were impending and there
was obviously some angst about whether or not it was the role of the ROD
Working Group to fully participate in the topic of community engagement;
some scientists thought that an advisory role in terms of the science was
a more appropriate role, leaving the community engagement to "others." I
checked in on this offline by informally participating in some of the activities that were being planned by those at the Institute and contacting those
who were working with, and members of, the native Hawaiian community. I
did this check-in with Merrie Monarch Festival coordinators by volunteering
and assisting with the Pua'ena'ena Ceremony, specifically helping to build
the stone-lined pit for the fire of Ke Ahi O Hi'iaka. This was my way of learning about, and also demonstrating my personal support of, the ceremony,
with respect and understanding of the endeavor. I also got a chance through
this check-in to meet those who were going to be benefiting from the decisions of the scientists on the ROD Working Group. As alluded to, or actually
dealt with, in a number of the previous chapters, there are often cultural,
societal, or other barriers to building fully trusting relationships among all
members of a work team, where scientists are involved. My experience in
Hawai'i was that a personal commitment to spend time on an issue, away
from work, is a tangible way to demonstrate that one is a trustworthy partner in the efforts, and this type of effort can help to break down barriers and
transform the dynamic of an otherwise overly science-oriented team, once
some relationship building steps are completed. So, again, in the example of
the ROD Working Group (and there were many other similar connections

like this, involving many other working group members), actions speak a thousand words in terms of building trust. The lack of this type of organically built trust is commonly what can lead to a lack of participation from community members, and conversely, frustration on the part of scientists who wonder why their information is not being used by the communities they are endeavoring to serve. Ultimately, in the ROD Working Group example, scientists were convinced that they could and should play a more active role in supporting the Merrie Monarch activities, simply by seeing that there was an appreciation for any of their involvement in the ceremonies, or outreach, such as participating in public service announcement development, ceremony participation, and consultations with the organizers of the events. I would submit that establishing any type of organically developed trust, early and often, leads to naturally occurring collaborative bonds, and a subsequent transformation of circumstances that all are aiming for. Once this initial and very critical step of building trust is on a reasonably good footing, there are so many more opportunities for authentic collaboration and transformation, which can then be explored further.

In the Federated States of Micronesia (FSM), there is enough trust among the people of Yap Island, through decades of experience and accomplishment together, to allow for an expanding collaboration among not just the people of FSM but also the U.S. Forest Service, Queen's University of Charlotte, The Nature Conservancy, Conservation International, the Micronesia Conservation Trust, and others. Yet, there is a need for additional technical and scientific expertise, which can be an opportunity in such situations, as this need helps to catalyze new collaborative linkages with those who have the expertise.

We can see this phenomenon in a number of the previous chapters, when a team's particular capabilities with regard to developing a collaborative approach are enhanced tremendously by bringing to bear the key types of technical and/or scientific expertise that is needed by the community we are collectively attempting to serve. This approach is one of several themes in Chapter 7, which demonstrates the very focused intent of the people of Cambodia, who have the same ecological endpoints of focus as the people of Việt Nam (mangrove restoration), but with a more deliberate plan of the government to restore and protect impacted ecosystems for the sake of their peoples' subsistence. The restoration and protection land ethic in Cambodia has been emphasized since the 1990s, and the amount of science-based approaches applied in the region has had an impact on driving the community-based management of coastal ecosystems. It is this work of the communities that has protected and rehabilitated mangroves throughout the region, and there is a solid basis of ecological knowledge supporting these efforts, which is capitalized upon by the local communities as well. Their practical approach to collaborative conservation in Cambodia is also bolstered by a reliable funding source that fuels the work, and in Cambodia, the projects are driven and funded in part by a number of innovative partners, namely, the International Development Research Centre; the

Australian People for Health, Education and Development Abroad; the Culture and Environment Preservation Association; and Mangroves for the Future. These partnerships did not just materialize spontaneously, they were incentivized by the funding of communities for the purposes of planting and protecting mangroves, allowing for practical work like community patrols to reduce mangrove forest degradation. Providing alternative fuels is also a great example of providing for the needs of people in Cambodia, in a way that is an environmentally acceptable alternative to harvesting mangroves, and directly serving the basic human needs for (in this case) fuel. This example also provides an important lesson, in that an important component of all successful partnerships is the fundamental respect and understanding that everyone requires the fundamentals of life. Finding alternatives that address these basic human needs is the challenge that is emerging in the early 21st century, and more collaborative approaches will certainly continue into the future, for the scientific and technological communities to address accordingly. To stimulate the financial side of the environmental solutions, as in Việt Nam, the promise of ecotourism is providing new options for environmental practices that are more sustainable than in the past, and science and technology can support these possibilities well, where this is an acceptable option. It should be noted that ecotourism is not always an option for communities, given certain societal constraints, concerns about ecologically sensitive areas, and (in some circumstances) incompatibility with community norms or traditions.

One of the key themes that you likely became aware of while reading this book is that the projects described outlined an inordinately large amount of time invested in building relationships, which ultimately resulted in the fruits of the successful collaborative work described, such as in Yap. Yap is a fantastic example of long-term relationship-building for successful environmental project management. As early as the 1970s, the involvement of an NGO in environmental work on Yap Island was established, i.e., the Yap Institute of Natural Science (YINS). This was a very intentional and practical approach, where the YINS was dedicated to the collection of ethnobiological, natural history, adaptive technological, and ecodevelopmental information. This naturally led to a more formal design for that partnership. An equivalent partnership in one of your project arenas may or may not be as formal or large as an institute like YINS, and could also simply be as informal as a partnership or consortium that is not chartered or otherwise formally structured—this was (and is) the case for the ROD Working Group (Chapter 1), which is an informal "consortium of the willing." Either way, it is always important to be looking out for these potential partnerships in your project arenas. One of the keys to the work in Yap with YINS is the local and familial ties of two of the project leads (i.e., authors Ruegorong and Falanruw), who reside on Yap and have long-standing familial ties to the communities being served. Both leads have utilized their familial and professional relationships there to such a degree that they are able to serve as a bridge, and indeed a guide, for others to span any possible gaps between the science, local traditions, technology, and societal norms. This is analogous in other situations with other project leads (perhaps you), who may live within the communities that are being served by scientists' work or by other expertise and knowledge. It is instructive to note in a number of the global stories of collaboration in the previous chapters that the more integrated into the daily life of the place a project lead is, the more likely it is that that they will have the added benefit of understanding the cultural, societal, and traditional nuances of the place, and then use that knowledge to more adeptly integrate traditional values into the science of the project place.

Another very good example that depicts trust being achieved, and leveraged to operationalize the science behind ecosystem restoration, is in Việt Nam (Chapter 7). A very mixed complement of approaches to community involvement is demonstrated in Việt Nam, which has been facing extreme environmental degradation in coastal areas since the 1960s, specifically in terms of mangrove loss along the coastal areas (i.e., an 82% loss in 30 years), and which has caused tremendous socioecological impacts upon the people of Việt Nam. Accordingly, this important topic is something that the current society in Việt Nam is endeavoring to amend. Because of the high priority of mangrove restoration and conservation by the government and the people of Việt Nam, areal coverage of mangroves has increased tremendously, due in large part to the establishment of collaboratives and through the direct participation of people who live in or near mangroves. This organic, local approach to ecosystem restoration and conservation has multiple positive impacts on the ecosystem, as well as the people involved in the collaboration.

A key factor for success in northern Việt Nam is effective community-based projects, where people from local communities are fully responsible for preparing and caring for mangroves, including everything from the planting, to protection, to long-term maintenance, and to social awareness (i.e., communication) of why these projects are important to everyone. The particular sociopolitical structure of Việt Nam lends itself to a people-based design, in some areas of Việt Nam, in that the government assigns the management of mangrove forests to the Provincial People's Committee of each commune in the area of restoration. The tiered design of this structure lends itself to a true community-based approach, as the responsibilities for the restoration work is apportioned across the various levels of responsibility, among communes and village communities, from the local government, police, military units, veterans, and village heads. These commune-based management models demonstrate the power and efficacy of solid community-based management of ecosystems in certain localities. In southern Việt Nam, the official policy is an opposite approach, where communities are not supposed to be involved according to the Civil Code. In reality, large areas of forests, including protected forests, are allocated to local communities for management, with the northern coastal region protected by women's or youth organizations. Aside from the administrative rules, in Việt Nam, the reality is that communities live in or near mangrove forests, such that their traditions are an integral part of everyday conservation work, furthering the integration of people and the land for generations to come. As with the inter-twining of society and nature in Việt Nam, we see the same ties of tradition and appreciation for wetlands as we did in the marshes of Iraq (Chapter 2), where people living in the marshes tends to lead to an increased presence and success of restoration of those projects.

Many of the areas that a community is responsible for in Việt Nam are mod-est in size and are funded through grant programs that enable a commune to take charge of their mangroves, to establish their own management structure so that they can be optimally responsible for the management and protection of mangrove forests assigned to them. The approaches in Việt Nam demonstrate how co-management is a good method in situations where the state would like to maintain some control over ecosystems but release responsibility for certain types of protection to local residents. This approach has been useful in Việt Nam, changing the mentality of residents from one of enforcement of banned access to a shared investment model where all who live in the area can help to protect the mangroves and control access for agreed-upon reasons. Through this shared interest approach, a foray into sustainable aquaculture for shrimp and ecotourism is being explored in Việt Nam, changing the focus from the prohibition of using the mangroves to a shared-values approach to developing sustainable uses of the mangroves, and it seems to be working.

When it comes to developing countries that are on the rise, Việt Nam is among a handful that stand out in terms of the vitality of their relatively (demographically dominant) young population, and their interest in taking approaches to environmental restoration that are environmentally, economi-cally, and socially sustainable. The people of Việt Nam, and their interna-tional partners, are working to make community-based management a meaningful balance between use and the perpetuation of ecological goods and services, demonstrating the operational side of the societal dimensions of the environmental sciences.

Another common theme of the many stories that bears on the topic of trust in the previous chapters is "a clash of cultures," which can epitomize interac-tions of many types, given the diversity and connectivity of our global soci-eties these days. When working to achieve environmental goals there is no reason why these 'trust issue' phenomena would be any different than with the other circumstances in which we see such conflicts arise, regardless of the type and location of the work. Many of these conflicts are simple mis-understandings, given differing cultural norms and traditions. When trust has been established, however, and operational goals are well laid out, these differing cultural and societal elements can even be used to build new path-ways of understanding, and thus accomplishing shared goals by using the diversity we bring to the task. In this way, traditions can be maintained and learning can be achieved, while accomplishing the goals of an environmen-tal project. The work on Yap Island (Chapter 4) demonstrates this approach wonderfully with great clarity, by acknowledging the past, the present, and the future for those who live in the Pacific. The reality of people's traditions and values lays out a potential problem that faces any and all environmental practitioners, leading us to a straightforward challenge question for all such collaborative teams: "How can we solve an environmental problem that exists in the modernized world, while honoring the traditions of the past?" We see that each of the practitioners in all of the chapters of this book grappled with this question, accomplishing this balance in a variety of ways.

Although many of the previous chapters describe a wide variety of circumstances that one might encounter with regard to the intentional integration of science, culture, and traditions, it is important to note that not all forms of integration require external forces in order to be a success, and indeed their success may come from (culturally and/or societally) internal design models. Such an internally designed structure occurs within a community, unto itself, and one where the needs and the solutions are endemic and part of the resident culture and/or community. A clear example of the phenomenon of this internal (i.e., organic and "endemic") process is the integration of science and culture described in Chapter 5, which outlines an efficient and seamless example of cultural norms driving science, and science solving a cultural need, i.e., a design that is integrated relatively effortlessly, and thoroughly, for those of the Islamic Faith. Indigenous knowledge and skills, namely, food, agriculture, and biological sciences, are used so that food products are culturally appropriate, such as it pertains to meat products and fermented foods or intoxicants. The critical importance of ensuring the authenticity of food products that comply with the laws of God is a primary need for those of faith in a number of religions, including Islam, Hinduism, Buddhism, Judaism, and others, and the practical aspects of this societal need is that worker protection, remediation, and disposal methods are thus guided by the scientific information attained through the analysis of foods. The science in Chapter 5 reinforces faith, and the outcome includes increasing public trust in the sourcing and purity of consumables, which is of paramount importance to a large proportion of the population of all societies worldwide, i.e., the global community of practice within Islam. Somewhat due to the full integration of science within this "global community of practice," Chapter 5 stands out among all of those in this book for its relative de-emphasis on named partnerships, and with little self-analysis of the collaborative process by which the food, agriculture, and biological sciences are utilized with contemporary technology to meet a fundamental societal (and indeed, cultural) need. Chapter 5 illustrates the seamlessness that can occur between science and society/culture, if both the technical and the societal dimensions are mature and connected, and focused on operational results, which is the case for their work that is based out of Malaysia, and with tremendous global influence. Chapter 5 serves to demonstrate to us all that once we have a solid integration of trust, knowledge, and coordination, the ties between the sciences and community needs can be very powerful and have far-reaching effects for the community(ies) involved, in both a geographically based (i.e., place-based) manner, or in a more general "community of practice" sense, on up to the global scale of influencing people's needs.

8.3 Leverage Your "Failures," "Crises," and "Miscommunications" for Success

You may have noticed that not all of the chapters in this book were full of "100% success stories," with perfectly happy endings, although most of them did result in some fairly successful outcomes, with others resulting in opportunities for further progress. As mentioned in Chapter 1, the challenges that are encountered when integrating the environmental sciences into a "societal needs construct" can be daunting, and often frustrating because near-term successes can be elusive. If you have ever been frustrated by these types of challenges, reading Chapter 2 should have made you feel in good company, and perhaps the level of difficulty in Iraq cured you of complaining about your particular environmental project woes. The impediments to success encountered in Iraq, postwar, are so overwhelmingly large in scope and number, one might consider giving up. However, those who live in Iraq have no such luxury and so their struggle continues every day to improve living conditions, and indeed what is at stake in Iraq are the very foundations of well-being (healthful air, water, and food), which we all must strive for. To this end, Iraqis are tireless in their pursuit of improving environmental conditions for their country, and they certainly epitomize the concept of embracing "failures" and learning from them when it comes to tackling current environmental problems. Through the process of rebuilding Iraq and improving environmental conditions there, the international community has also stepped forward to address the extreme needs in Iraq, bringing funding to bear on the issues, matching the passion and fortitude of Iraqis. The United Nations and the Iraq Government have evidently decided that this work is a priority, and both have played a very large role, particularly in funding and providing relevant scientific expertise, along with the indigenous scientific expertise that exists in-country. The circumstances of the country, now, finally postwar, have also attracted a number of expatriates who regularly return to Iraq to assist, such as does the author of Chapter 2, an accomplished remote sensing scientist in the United States. One might imagine any other environmental issue that might arise in another location around the world, or set of circumstances, but it is difficult to envision a place or circumstance that is in circumstances as dire as those faced in Iraq. Nevertheless, Iraqis are working hard to mitigate environmental pollution and to make lives better in their war-torn country, which has left a legacy that will likely take a large portion of this century to rectify.

As alluded to in Section 8.1, the two major emphases in all of the stories of this book were "people and their needs" and (secondarily) the "science that serves the needs of people." This relative emphasis was very intentional on the authors' parts, lending to the unique presentation style of environmental science in this book. By the same token, the focus of this book has not been solely focused upon: (A_1) "decision makers," nor has it been on (B_1) "science delivery," but rather: (A_2) across the full spectrum of interested parties in whatever the particular environmental issues are at a locale, from agency and community leaders to members of the public, and (B_2) the sharing of knowledge. Note the subtle differences in the contrasted topics (i.e., compare both A's and then both B's) in the previous sentence; the contrast of A's and B's above is made to drive home the point that most of the successes of the projects described in this book were due to a certain type of flexibility and openness on the part of both scientists and nonscientists, with the general theme of bidirectional knowledge sharing. This approach is epitomized by the story told in Chapter 3, where the people of the Ozark region of the United States were in control of their own destiny, during which time they drew upon the available science (such as the U.S. Environmental Protection Agency [USEPA] watershed-based research conducted there in the early 2000s) in such a way as to provide information and energy to their long-held values, for their beloved river and landscape. The USEPA study propelled some portion of the watershed planning group work, but it was not entirely responsible for the energy surrounding the public's involvement and excitement about caring for their watershed (Chapter 3).

One additional key learning from the work in the Ozarks is that the impact of any scientific or technical endeavor can far outlast the length of the scientific study. For this reason, it is important, and recommended, for environmental scientists to realize the long-lasting and far-reaching effects of their work upon communities, which impacts people far beyond the analytical work that they conduct and share, even if the research project is just for a few years. It may also be of tremendous help to revisit such communities long after the science analyses have been completed (such as the authors did in Chapter 3), and (re)discover what the true societal impacts of your science team was and is. We (my Chapter 3 coauthors and I) learned from the Ozarks work that, in very special and unforeseen ways, the places where we conducted the scientific work will never be the same because of our endeavors, and there is much to be proud of in terms of the positive and lasting ways we affected the lives of people in the Ozarks.

8.4 Listen Much More, and Talk a Little Less

As the Greek Philosopher Diogenes reasoned, "We have two ears and one tongue so that we would listen more and talk less," which is quite true to form for Diogenes in his bluntness, explaining that we must endeavor to stop, take pause, and listen to others in order to fully understand the world around us, and indeed the needs of others. No matter what role you play in environmental projects, it is always more valuable to listen than to "profess," and those around us always have something of value to add to the conversation, even if we disagree with their assertions. A contemporary example of this circumstance comes from a colleague of mine who is a professionally trained conflict resolution specialist, whose mantra has always been, "words are just information, nothing else," alluding to how our ego can cause us to take words as criticisms, thus causing us to 'not hear' what is said." Indeed, my colleague's mantra can serve us all as a gift in situations where we are having a conversation with someone on a "collaborative team" who may not share our values or ideals. The information you receive from such a person may even tell you to cease working or communicating with them, but nevertheless, the words are good information, and thus truly a gift to receive and appreciate. As scientists, especially, we should be particularly able to approach new information in this manner, i.e., to value the words of other people or the views of communities we are endeavoring to serve, as we would value all types of information as part of our training as scientists. Accordingly, scientists are trained to gather all available information (such as the commentary of our partners), look at the information objectively (without interjecting our personal opinions), and then utilize this information to its fullest to build new models of understanding of the world around us. This approach is a precept of the "scientific method," which can be adapted to listening and learning more effectively with our communities and partners. Listening to the needs of people, and learning from that experience, is a primary part of public service, by definition, too, which may apply to you if you are a civil servant, or if not is also a common tenet of our civil society and serves us particularly well if we are scientists, in that scientists depend upon "sound and current data" (in this example, from partners) for their work to proceed well. So, if we as scientists cease to listen to those who are providing us with "sound and current data" about their needs, we will miss important clues about what it is we need to understand in order to effectively address their problems. After all, a key goal of most environmental scientists that deal with issues that are in the public realm is to effectively address applied, societally-relevant topics.

Overall, this book takes a very simple approach to a very complex issue, and one that is very similar to the approach that I and others have employed in our careers as an environmental scientists: (1) have an honest and simple dialog with people; (2) focus on listening, primarily; and (3) respond to the needs of those that speak with you, both honestly and completely. One of the best examples I have seen of this approach in this book, in a structured manner, is described in Chapter 6, where the intentional use of stakeholder "appreciative inquiry" and involvement was used to redress an earlier session that was not as well received, and which led to a much better outcome when the dialog approach was changed to focus more on listening to the stakeholders in the Midwestern Region of the United States. The "appreciative inquiry" approach led to much more genuine (and useful) stakeholder comments and concerns about the future landscape scenarios being proposed by USEPA, in terms of their lives and concerns.

The approach taken in the Midwestern Region of the United States is also very much in line with the fundamentals of the concept of transdisciplinarianism, the body of thought that the inclusion of all potential stakeholders (from the very beginning of a project) enhances the conversations that occur about the issues involved in a project, thus ensuring that all necessary voices and perspectives are included in the outcomes of the project (Rosenfield 1992). In the "Future Midwestern Landscapes" work, the "change drivers of concern" were ultimately developed from such a transdisciplinary group of individuals and stakeholders, epitomizing the very best of what might be a structured process for bringing in all necessary perspectives, leading to a successful project outcome/plan.

8.5 Epilogue

This book is a first of its kind, which is to simply say that we, environmental professionals and people/communities with environmental issues to solve, have much more to discuss and thousands of stories to tell—all of us. Community members, academics, environmentalists, scientists, natural resources managers, and others have a responsibility to analyze and share the societal dimensions of their work more, and although there has been a historical bias against doing just that, it is clear that the trend is for more holistic approaches to solving environmental problems in the future and integrating the many "ways of knowing" the issues involved. It is no secret that the "traditional sciences" have a custom of giving short shrift to community needs, traditional values, and the modest (but critical) needs of people, but the most important thing to recognize is that this trend is changing quickly, and indeed, the trend is for much more inclusion of societal needs in all environmental science projects of the future. With a relatively recent emphasis on partnerships and collaboration, the world of contemporary environmental science, which, for much of the 20th century was focused mainly on "pure science" or "applied science," is now opening up to the broader public, worldwide, and becoming much more accessible to individual community members, including the innovative communication methods opened up by Internet-based interfaces and technology (such as described in Chapter 1). As a community of science, let us embrace this relatively new emergence of information sharing, just as we have embraced digital data over the past several decades, substantially replacing analogue techniques for accomplishing scientific data gathering and analysis. It is with a similar zeal and sophistication that we must now lean into building a digital community-based approach to the environmental sciences.

The new challenge for us all is to find fresh methods for operationally including societal dimensions into our work as environmental scientists, and it is with this volume of stories that you may find inspiration and practical guidance to do just that, offering encouragement, energy, and practical information for implementation to your particular environment projects. Consider referring back to the chapters in this book often, as they may spark innovative ideas for developing your projects and overcoming impediments that may arise. By digging deeper into a number of projects in the previous chapters, you will likely observe that the common themes ring true regardless of the project type, or the locale of the project. Please also use the examples in this book to motivate others on your team, and to prompt you to listen a bit differently to your team members and other community members, as you go about your work. Most importantly, observe how being in the "inquiry mode" with your team and community members can positively impact the dynamics of the group, and the project you are working on; this inquisitive approach is guaranteed to elicit a positive outcome for the project, and consequently the communities you are endeavoring to serve, utilizing the environmental sciences.

It is on these hopeful notes that we complete our journey of learning, through the global stories of collaboration and transformation. We welcome your feedback on these stories and indeed hope you might share yours with us; please feel free to send your stories of collaboration and transformation, along with any other comments or information to us, at: Stories.of.Collab.and.Trans@gmail.com, so that we might expand the conversation about successful approaches, effective solutions, and cross-pollinating ideas, for the purpose of meeting the goals and needs of science and those of society as a whole.

Appendix A
Welina ke aloha dearest Hula people of Hawai'i and the world

Our profound aloha to each and every—'Olohe, Lo'ea and Loiloi Hula, Kumu Hula, Po'o Pua'a, Alaka'i, Paepae, and Haumāna Hula! Aloha pū to the 'ohana, musicians, singers, chanters, lei makers, researchers, dresser people, seamstresses, and hair and makeup specialists who help us recreate and relive every hula, mele, and story in the best way.

My name is Kekuhi Kanae Kanahele Keali'ikanaka'oleohaililani. I am a granddaughter of a most beloved Tūtū and Kumu Hula who taught me, in every moment of my time with her, to BE IN LOVE with, and have a huge ALOHA for all of our 'ohana. She dedicated her time with me to the simple task of making sure I knew how important ALL of our 'ohana are.

When Grampa took us visiting with family around the island Gramma would say, "this is your 'ohana," then sit down to wala'au or talk story. When we visited Hā'ena, Halema'uma'u or Punalu'u, she told me, "this is your 'ohana," and then told stories of those places. At the ocean when we fished, she would say, "this is your 'ohana," and offer a chant to the water, and the fish on our hooks. At the māla, she would say, "this is your 'ohana," and sing a song to the taro. *At the forest, she would pick liko from the 'Ōhi'a, hold it in her fingers, and say, "Kekuhi, this is your 'ohana," and we would give a chant of thanks.*

This memory has become an every moment practice—I continue to dedicate my life to dancing, chanting, singing, and teaching how to Aloha ALL of these 'ohana...like many of you do in big and little ways through HULA. I am writing because our hula 'ohana, the tree that is most used in our art form, the tree that is most responsible for making sure that we have water, the tree that is most used in the carving of ki'i, the tree that we can find on almost every landscape on our island, the tree that many of our bird people depend on, the tree that Hopoe and Hi'iaka made lehua lei from, the beloved 'ŌHI'A...the tree that my Gramma introduced to me as 'ohana—is being made sick by a fungus, *Ceratocystis fimbriata* (aka Rapid 'Ōhi'a Death or ROD), that lives both in the soil and in the tree. OUR 'Ōhi'a needs our attention, our awareness, and our aloha.

We don't know how this particular strain of fungus got here or how to treat it properly...yet. But, we do know some simple, but effective things that everyone can do to help prevent the spread of the fungus to healthy trees on Hawai'i island and to other islands.

My friends of the forest from around the island and I are working together to care for and find ways to heal our beloved 'Ōhi'a...the 'Ōhi'a deserve that—anyone who has ever been sick deserves that! Here are some simple preventative actions that we as Kumu, dancers, lei makers, and 'ohana can do:

- If you need to collect 'ōhi'a wood, avoid areas known to have the fungus, or areas that look to be sick from the fungus.
- Leave your lei and kūpe'e, that have liko and lehua in it, on Hawai'i island. When returning your lei and kūpe'e to our forest, return it to the same area that your liko and lehua came from. Or return your lei and kūpe'e to any of the forest areas identified on the attached map. In doing so, you are giving your mana to a forest that needs healing.

  For the 2016 Merrie Monarch Festival in Hilo, Hawai'i, the Hawai'i Island community is hosting the Pua'ena'ena Ceremony. This fire ceremony will provide a way for people to offer their kinolau, hakina, lei, and kūpe'e with thoughts of full recovery for our 'Ōhi'a to the fire of Ke Ahi O Hi'iaka (see details on invitation announcement).
- Brush clean the dirt from your shoes and any tools you used, and spray everything with 70% rubbing alcohol when entering and leaving the forest (visit the DOFAW office in Hilo to learn more, and for some supplies).
- Wash your vehicle, its tires and the undersides if you drive off-road. Keep the inside of your vehicle clean of dirt, and spray the floors of your vehicle with 70% rubbing alcohol.
- Notice and report 'Ōhi'a that are dying in any area that you go to pick, especially areas that are not already on the map of affected forests (contact the folks on the brochure).

Aside from all of those things above, as the hula community is well aware, our songs, hula, chants, prayers, and our thoughts of full recovery sent directly to our 'Ōhi'a community are just as powerful!

Ulu ka 'O-hi'a...a lau ka wai

Kekuhi

DOFAW IDENTIFIED HOʻIHOʻI FOREST RESERVE LOCATIONS

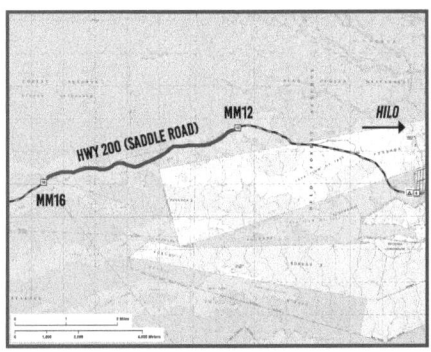

SADDLE ROAD

The Division of Forestry and Wildlife hoʻihoʻi zone (in red on map) located between mile markers 12 and 16. Safe pullouts are located at MM 12 on the right side of road, and at MM 16 on the left and right sides of road. Please hoʻihoʻi no more than 20ft in from forest edge.

STAINBACK HIGHWAY

The Division of Forestry and Wildlife hoʻihoʻi zone (in red on map) located three miles from Highway 11 (Kanoelehua) along Stainback Highway to North Kūlani Road. Please hoʻihoʻi no more than 20ft in from forest edge.

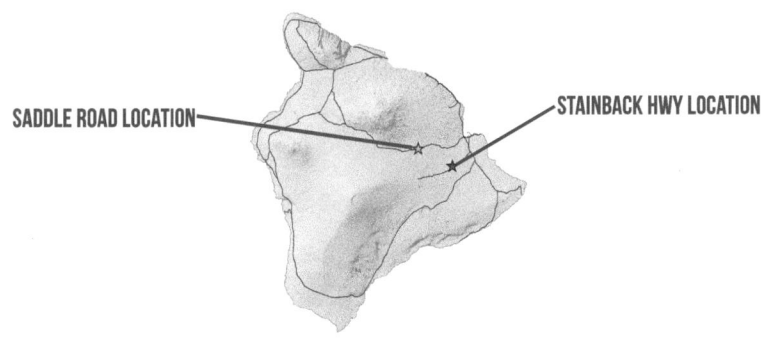

RAPID 'ŌHI'A **DEATH**

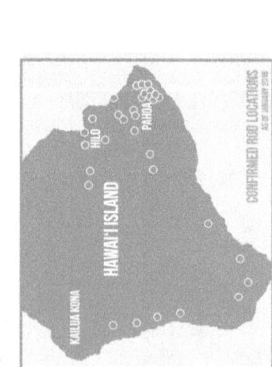

WHAT YOU CAN DO
TO HELP PREVENT THE
SPREAD

BROCHURE UPDATED FEBRUARY 2018

FOR THE LATEST INFORMATION, MAPS, AND
UPDATES ON RAPID 'ŌHI'A DEATH PLEASE VISIT:

www.rapidohiadeath.org
f www.facebook.com/rapidohiadeath

**IF YOU SUSPECT ROD IN YOUR AREA
PLEASE CONTACT:**

Dr. J.B. Friday
UH Cooperative Extension Service
Email: jbfriday@hawaii.edu
Phone: (808) 969-8254

Dr. Flint Hughes
USDA Forest Service
Email: fhughes@fs.fed.us
Phone: (808) 854-2617

Dr. Lisa Keith
USDA Agriculture Research Service
Email: Lisa.Keith@ars.usda.gov
Phone: (808) 959-4357

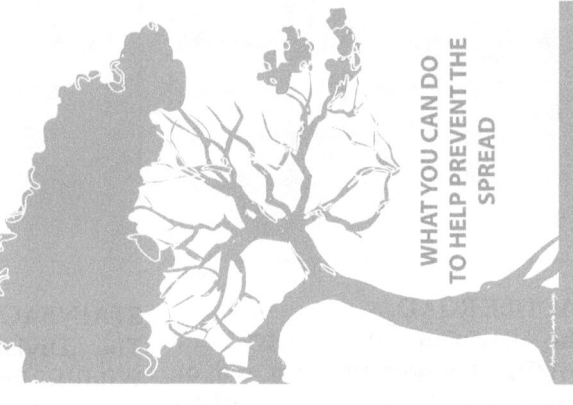

UH MANOA COLLEGE OF TROPICAL AGRICULTURE AND HUMAN RESOURCES
USDA AGRICULTURE RESEARCH SERVICE
USDA INSTITUTE OF PACIFIC ISLANDS FORESTRY
DEPARTMENT OF LAND AND NATURAL RESOURCES, DIVISION OF FORESTRY AND WILDLIFE

RAPID 'ŌHI'A DEATH A NEWLY IDENTIFIED DISEASE

A disease that is new to science and new to Hawai'i has killed hundreds of thousands of 'ōhi'a trees (*Metrosideros polymorpha*) across more than 34,000 acres of Hawai'i Island. Known as Rapid 'Ōhi'a Death (ROD), it is caused by a fungus called *Ceratocystis fimbriata*. We can all help minimize further spread, and buy time to help researchers find answers and potential treatments. New information is being uncovered almost every week.

This disease has the potential to kill 'ōhi'a trees statewide. The most important thing we can all do now is help **prevent ROD from spreading**.

'Ōhi'a lehua is the backbone of Hawai'i's native forests and watersheds which are our source of fresh water in these islands. 'Ōhi'a trees cover more than 1 million acres statewide—they are perhaps the most important tree in Hawai'i.

HAWAI'I ISLAND

KAILUA KONA

HILO

PĀHOA

CONFIRMED ROD LOCATIONS
AS OF JANUARY 2018

5 THINGS YOU CAN DO

1 DON'T MOVE 'ŌHI'A

Do not move 'ōhi'a wood, firewood or posts, especially from an area known to have ROD. If you don't know where the wood is from, don't move it.

2 DON'T TRANSPORT 'ŌHI'A INTER-ISLAND

Comply with the new quarantine rule to help prevent ROD from spreading. Don't move 'ōhi'a plants, wood, or other 'ōhi'a plant parts inter-island without a permit.

3 CLEAN YOUR TOOLS

Use only these proven cleaning methods—other methods have been tested and they don't kill the fungus. Tools used for cutting 'ōhi'a trees (especially infected ones) should be cleaned with 70% rubbing alcohol.

4 CLEAN YOUR GEAR

Clean your shoes, and clothing. Decontaminate shoes by dipping in 70% rubbing alcohol to kill the ROD fungus. Other gear can also be sprayed with 70% rubbing alcohol. Wash clothing in hot water and detergent.

5 WASH YOUR VEHICLE

Wash the tires and undercarriage of your vehicles with detergent and remove all soil or mud, especially after traveling from an area with ROD and/or if you have traveled off-road.

NEW QUARANTINE RULE

Because this disease could have devastating impacts on Hawai'i's 'ōhi'a forests, and people could accidentally spread it, the Hawai'i Department of Agriculture passed a new quarantine rule that prohibits interisland movement except by permit of all 'ōhi'a plant or plant parts including:

- Logs, wood, leaves, twigs, flowers, seeds, stems, cuttings.
- Untreated wood, green waste, mulch, sawdust, wood chips and frass (wood dust from boring beetles).

The movement of soil is also prohibited except by permit.

All these materials could carry the fungus and spread the disease. The fungus can stay viable for over a year.

To apply for permits to move 'ōhi'a products that can be demonstrated to be free of disease or soil, please visit: **HDOA.HAWAII.GOV**

Symptoms of the Disease

- Crowns of 'ōhi'a trees that appear healthy turn yellowish or brown within days to weeks; dead leaves remain on branches for some time.
- All ages of 'ōhi'a trees can be affected and can have symptoms of browning of branches and/or leaves.
- If a tree with ROD is cut down, or a section of the tree is removed, the fungus shows up as dark staining in the sapwood along the outer edge, and there may be an over-ripe fruit-like odor.
- Trees within a given stand die in a haphazard pattern; the disease does not appear to radiate out directly from infected or dead trees.

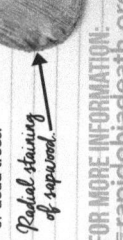

Radial staining of sapwood.

FOR MORE INFORMATION:
rapidohiadeath.org
USDA

Literature Cited

Chapter 1

Aradhya, K. M., D. Mueller-Dombois, and T. A. Ranker. 1993. Genetic structure and differentiation in Metrosideros polymorpha (Myrtaceae) along altitudinal gradients in Maui, Hawaii. *Genetical Research* 61: 159–170.

Cordell, S., G. Goldstein, F. C. Meinzer, and P. M. Vitousek. 2001. Regulation of leaf life-span and nutrient-use efficiency of Metrosideros polymorpha trees at two extremes of a long chronosequence in Hawaii. *Oecologia* 127: 198–206.

Cordell, S., G. Goldstein, P. J. Melcher, and F. C. Meinzer. 2000. Photosynthesis and freezing avoidance in Ohia (Metrosideros polymorpha) at treeline in Hawaii. *Arctic Antarctic & Alpine Research* 32(4): 381–387.

Cordell, S., G. Goldstein, D. Mueller-Dombois, D. Webb, and P. M. Vitousek. 1998. Physiological and morphological variation in Metrosideros polymorpha, a dominant Hawaiian tree species, along an altitudinal gradient: The role of phenotypic plasticity. *Oecologia* 113: 188–196.

Cornwell, W. K., B. Bhaskar, L. Sack, S. Cordell, and C. K. Lunch. 2007. Adjustment of structure and function of Hawaiian Metrosideros polymorpha at high vs. low precipitation. *Functional Ecology* 21: 1063–1071.

Crawford, N. G., C. Hagen, H. F. Sahli, E. A. Stacy, and T. C. Glenn. 2008. Fifteen polymorphic microsatellite DNA loci from Hawaii's Metrosideros polymorpha (Myrtaceae: Myrtales), a model species for ecology and evolution. *Molecular Ecology Resources* 8: 308–310.

Ellison, A. M., M. S. Bank, B. D. Clinton, E. A. Colburn, K. Elliott, C. R. Ford, D. R. Foster, B. D. Kloeppel, J. D. Knoepp, G. M. Lovett, J. Mohan, D. A. Orwig, N. L. Rodenhouse, W. V. Sobczak, K. A. Stinson, J. K. Stone, C. M. Swan, J. Thompson, B. Von Holle, and J. R. Webster. 2005. Loss of foundation species: Consequences for the structure and dynamics of forested ecosystems. *Frontiers in Ecology and the Environment* 9: 479–486.

Fisher, J. B., G. Goldstein, T. J. Jones, and S. Cordell. 2007. Wood vessel diameter is related to elevation and genotype in the Hawaiian tree Metrosideros polymorpha (Myrtaceae). *American Journal of Botany* 94(5): 709–715.

Gruner, D. S. 2004. Arthropods from 'ōhi'a lehua (Myrtaceae: Metrosideros polymorpha), with new records for the Hawaiian Islands. *Bishop Museum Occasional Paper* 78: 33–52.

Harbaugh, D. T., W. L. Wagner, D. M. Percy, H. F. James, and R. C. Fleischer. 2009. Genetic structure of the polymorphic Metrosideros (Myrtaceae) complex in the Hawaiian Islands using nuclear microsatellite data. *PLoS One* 4(3): e4698.

Izuno, A., M. Hatakeyama, T. Nishiyama, I. Tamaki, R. Shimizu-Inatsugi, R. Sasaki, K. Kentaro, K. Shimizu, and Y. Isagi. 2016. Genome sequencing of Metrosideros polymorpha (Myrtaceae), a dominant species in various habitats in the Hawaiian Islands with remarkable phenotypic variations. *Journal of Plant Research* 129(4): 727–736.

James, S. A., C. F. Puttock, S. Cordell, and R. P. Adams. 2004. Morphological and genetic variation within Metrosideros polymorpha (Myrtaceae) populations from different habitats on Hawai'i. *New Zealand Journal of Botany* 42: 263–270.

Keith, L. M., R. F. Hughes, L. S. Sugiyama, W. P. Heller, B. C. Bushe, and J. B. Friday. 2015. First report of Ceratocystis wilt on 'Ōhi'a (Metrosideros polymorpha). *Plant Disease* 99(9): 1276.

Loope, L. L. and T. W. Giambelluca. 1998. Vulnerability of island tropical montane cloud forests to climate change, with special reference to East Maui, Hawaii. *Climatic Change* 39: 503–517.

Mortenson, L., R. F. Hughes, J. B. Friday, L. M. Keith, J. M. Barbosa, N. J. Friday, Z. Liu, and T. G. Sowards. 2016. Assessing spatial distribution, stand impacts and rate of Ceratocystis fimbriata induced 'Ōhi'a (Metrosideros polymorpha) mortality in a tropical wet forest, Hawai'i Island, USA. *Forest Ecology and Management* 377(1): 83–92.

Percy, D. M., A. M. Garver, W. L. Wagner, H. F. James, C. W. Cunningham, S. E. Miller, and R. C. Fleischer. 2008. Progressive island colonization and ancient origin of Hawaiian Metrosideros (Myrtaceae). *Proceedings of the Royal Society of London, Series B* 275: 1479–1490.

Pratt, T. K., C. T. Atkinson, P. C. Banko, J. D. Jacobi, B. L. Woodworth, and L. A. Mehrhoff. 2009. Chapter 24, can hawaiian forest birds be saved? In T. K. Pratt, C. T. Atkinson, P. C. Banko, J. D. Jacobi, and B. L. Woodworth (eds.). *Conservation Biology of Hawaiian Forest Birds: Implications for Island Avifauna*. New Haven, CT: Yale University Press.

Rosenfield, P. L. 1992. The potential of transdisciplinary research for sustaining and extending linkages between the health and social sciences. *Social Science & Medicine* 35(11): 1343–1357.

UNEP. 2005. *Assessment of Environmental "Hot Spots" in Iraq*. Switzerland: United Nations Environment Programme.

Vitousek, P. 2004. *Nutrient Cycling and Limitation: Hawai'i as a Model System*. Princeton, NJ: Princeton University Press.

Zimmerman, E. C. 1948. *Insects of Hawaii: Volume 4*. Honolulu: University of Hawaii Press.

Chapter 2

Al-Ansari, N. A. 2013. Management of water supplies in Iraq: Problems and prognosis. *Engineering* 5: 667–684.

Al-Mudaffar, N., K. P. Goodwin, B. A. Mahdi, and M. L. Stevens. 2016. Effects of Mesopotamian marsh (Iraq) desiccation on the cultural knowledge and livelihood of Marsh Arab women. *Ecosystem Health and Sustainability* 2(3): e01207.

Anderson, J. W. 2004. An analysis of a dust storm impacting operation Iraqi freedom, 25–27 March 2003. *Master's Thesis*. Naval Post Graduate School, Monterey, CA, USA.

Cave, D. and A. Fadam. 2007. Iraq insurgents employ chlorine in bomb attacks. *New York Times*. www.nytimes.com/2007/02/22/world/middleeast/22iraq.html checked September 15, 2018.

Chenoweth, J., P. Hadjinicolaou, A. Bruggeman, J. Lelieveld, A. Levin, M. A. Lange, E. Xoplaki, and M. Hadjikakou. 2011. Impact of climate change on the water resources of the Eastern Mediterranean and Middle East Region: Modeled 21st century changes and implications. *Water Resources Research* 47: 1–8.

COMET (Program). 2003. Forecasting dust storms, Univ. Corp. for Atmos. Research, on-line training program. www.meted.ucar.edu/training_module.php?id=782#. W6uwsf6WxLM checked September 15, 2018.

Environment News Service. 2014. UN to help Iraq with environmental restoration. http://ens-newswire.com/2014/01/26/un-to-help-iraq-with-environmental-restoration checked September 15, 2018.

Faa, A., C. Gerosa, D. Fanni, G. Floris, P. V. Eyken, J. I. Lachowicz, and V. M. Nurchi. 2018. Depleted uranium and human health. *Current Medicinal Chemistry* 25(1): 49–64.

FAO. 2015. Country fact sheet, AQUASTAT survey Iraq. Food and Agriculture Organization of the United Nations. www.fao.org/nr/water/aquastat/data/cf/readPdf.html?f=CF_IRQ_en.pdf checked September 15, 2018.

Iraq Ministry of the Environment. 2014. The state of environment and outlook report (in Arabic). http://ens-newswire.com/2014/01/26/un-to-help-iraq-with-environmental-restoration/ checked February 06, 2019.

Kubba, S. 2011. *The Iraqi Marshlands and the Marsh Arabs: The Ma'dan, Their Cultures and the Environment*. Ithaca, NY: Ithaca Press.

Schlanger, Z. 2018. ISIL is lighting oil wells on fire as they retreat, and no one is paying attention. *QUARTZ*. https://qz.com/1182389/isil-is-lighting-oil-wells-on-fire-as-they-retreat-and-no-one-is-paying-attention/ checked September 21, 2018.

UNEP. 2003. *Environment in Iraq: UNEP Progress Report*. Geneva: United Nations Environment Programme.

UNEP. 2005. *Assessment of Environmental "Hot Spots" in Iraq*. Geneva: United Nations Environment Programme.

UNEP. 2007. *UNEP in Iraq: Post-Conflict Assessment, Clean-up and Reconstruction*. Nairobi: United Nations Environment Programme.

UNEP. 2011. *Keeping Track of Our Changing Environment: From Rio to Rio+20 (1992–2012)*. Nairobi: United Nations Environment Programme.

UNEP. 2016. *Iraq's Marsh Arabs More Optimistic after World Heritage Status*. Iraq: United Nations Environment Programme.

UNEP. 2017. *The National Environmental Strategy and Action Plan for Iraq (2013–2017)*. Iraq: United Nations Environment Programme.

WHO (Regional Office for the Eastern Mediterranean). 2017. *Iraq Health Profile 2015*. Geneva: Regional Office for the Eastern Mediterranean, World Health Organization.

Zwijnenburg, W. 2015. Iraq's continuing struggle with conflict pollution. *Peace Insight*. www.peaceinsight.org/blog/2015/03/iraqs-continuing-struggle-conflict-pollution/ checked September 15, 2018.

Chapter 3

Beaver Water District. 2008. *Beaver Lake and its Watershed—2008*. Fayetteville, AR: Arkansas Water Resources Center.

Campbell, B. C. 2010. Closest to everlastin: Ozark agricultural biodiversity and subsistence traditions. *Southern Spaces*. https://southernspaces.org/2010/closest-everlastin-ozark-agricultural-biodiversity-and-subsistence-traditions checked September 15, 2018.

Lopez, R. D. and R. C. Frohn. 2018. *Remote Sensing for Landscape Ecology Monitoring, Modeling, and Assessment of Ecosystems*, 2nd edition. Boca Raton, FL: CRC Press.

Masters, R. D. 1998. *Fortune Is a River*. New York: Penguin Group.

Ozarks Water Watch. 2018. *Map Graphics on Watershed Map of the Upper White River Basin*. Lansing, MI: Missouri Department of Natural Resources.

Postel, S. and B. Richter. 2003. *Rivers for Life-Managing Water for People and Nature*. Washington, DC: Island Press.

Schoolcraft, H. R. 1821. *Journal of a Tour into the Interior of Missouri and Arkansas*. London: Sir R. Phillips and Company.

Schoolcraft, H. R. 1996. *Rude Pursuits and Rugged Peaks: Schoolcraft's Ozark journal 1818–1819*. With an Introduction, Maps, and Appendix by Milton D. Rafferty. Fayetteville, AR: University of Arkansas Press.

Chapter 4

Ananthaswamy, A. 2012. Projections of sea level rise are vast underestimates. *New Scientist*. www.newscientist.com/ checked February 06, 2019.

Australian Bureau of Meteorology and CSIRO. 2014. *Climate Variability, Extremes and Change in the Western Tropical Pacific: New Science and Updated Country Reports*. Pacific-Australia Climate Change Science and Adaptation Planning Program Technical Report. Melbourne, VIC: Australian Bureau of Meteorology and Commonwealth Scientific and Industrial Research Organisation.

Beck, T. 2013. *Principles of Ecological Landscape Design*. Washington, DC: Island Press.

Cairns, M. F. (ed.). 2007. *Voices from the Forest: Integrating Indigenous Knowledge into Sustainable Upland Farming*. London: Earthscan.

Cairns, M. F. (ed.). 2015. *Shifting Cultivation and Environmental Change: Indigenous People, Agriculture and Forest Conservation*. London: Taylor & Francis Press.

Central Intelligence Agency. 2017. Federated states of Micronesia. *The World Factbook*. www.cia.gov/library/publications/resources/the-world-factbook/geos/fm.html checked September 15, 2018.

Chief Executives of Micronesia. 2006. Declaration of commitment: 'The Micronesia Challenge'. http://www.fsmpio.fm/ checked February 06, 2019.

Chui, T. F. and J. P. Terry. 2015. Groundwater salinization on atoll islands after storm-surge flooding: Modelling the influence of central topographic depressions. *Water and Environment Journal* 29(3): 430–426.

Cole, T., M. Falanruw, C. D. MacLean, C. D. Whitesell, and A. H. Ambacher. 1987. *Vegetation of Republic of Palau*, Resource Bull. PSW-22. Berkeley, CA: USDA Forest Service.

Daily, G. and K. Ellison. 2002. *The New Economy of Nature*. Washington, DC: Island Press.

de Oca, J. 1893. La Isla de Yap. *Boletin de la Sociedad Geographica de Madrid* XXXIV(4–6): 251–179.

Denevan, W. M. and J. J. Parsons. 1967. Pre-columbian ridged fields. *Scientific American* 217(1): 92–100.

Denevan, W. M. 1970. Aboriginal drained-field cultivation in the Americas. *Science* 169(3946): 647–654.

Denevan, W. M. 1982. Hydraulic agriculture in the American tropics: Forms, measures and recent research, In K. V. Flannery (ed.). *Maya Subsistence*. New York: Academic Press.

Dodson, J. R. and M. Intoh. 1999. Prehistory and palaeoecology of Yap, federated states of Micronesia. *Quaternary International* 59(1): 17–26.

Donato, D. C., J. B. Kauffman, S. Kurnianto, M. Stidham, and D. Murdiyarso. 2011. Mangroves among the most carbon-rich forests in the tropics. *Nature Geoscience* 4: 293–297.

Donato, D. C., J. B. Kauffman, R. A. Mackenzie, A. Ainsworth, and A. Z. Pfleeger. 2012. Whole—island carbon stocks in the tropical Pacific: Implications for mangrove conservation and upland restoration. *Journal of Environmental Management* 97: 89–96.

Falanruw, M. V. C. 1979. *Environmental Education Handbook*. Noumea: South Pacific Commission Press.

Falanruw, M. V. C. 1985. People pressure: Management of limited resources on Yap, In W. McNeely (ed.). *The Role of Rotected Areas in Sustaining Society*. Washington, DC: Smithsonian Institution Press.

Falanruw, M. V. C., T. Cole, A. H. Ambacher, K. E. McDuffie, and J. E. Maka. 1987a. *Vegetation Survey of Moen, Dublon, Fefen and Eten, State of Truk, Federated States of Micronesia*, Resource Bulletin, PSW-20. Berkeley, CA: USDA Forest Service.

Falanruw, M. V. C., T. G. Cole, and C. D. Whitesell. 1987b. The vegetation of acid soils of Micronesia. *Proceedings of International Soil Management Workshop, Management and Utilization of Acid Soils of Oceania, Belau, February 2–6, 1987*. Agriculture Experiment Station, University of Guam, Mangilao, Guam.

Falanruw, M. V. C., C. D. Whitesell, T. G. Cole, C. D. MacLean, and A. H. Ambacher. 1987c. *Vegetation survey of Yap, Federated States of Micronesia*, Resource Bulletin, PSW-21. Berkeley, CA: USDA Forest Service.

Falanruw, M. V. C. 1988. On the status, reproductive biology and management of fruit bats of Yap. *Micronesia* 21(1–2): 39–51.

Falanruw, M. V. C. 1989a. *The Vegetation of Asuncion: A Volcanic Northern Mariana Island*, Resource Bulletin, PSW-28. Berkeley, CA: USDA Forest Service.

Falanruw, M. V. C. 1989b. *Yap Fruit Bat Management, Past, Present and Future Challenge*, vol. 18. Bethesda, MD: Transactions of the Western Section, the Wildlife Society.

Falanruw, M. V. C., T. Cole, and A. H. Ambacher. 1989. *Vegetation Survey of Rota, Tinian and Saipan, Commonwealth of the Northern Mariana Islands*, Resource Bulletin, PSW-27. Berkeley, CA: USDA Forest Service.

Falanruw, M. V. C. and C. J. Manmaw. 1990. *Protection of Flying Foxes On Yap Island. Symposium on Flying Foxes*. Honolulu: Bat Conservation International.

Falanruw, M. V. C. 1990a. Micronesian agroforestry: Evidence from the past, implications for the future, In B. Raynor and R. Bay (eds.). *Proceedings of the Workshop on Research Methodologies and Applications for Pacific Island Agroforestry (July 16–20). Kolonia, Pohnpei, Federated States of Micronesia*, Gen. Tech. Rep. PSW-GTR-140, Albany, CA: USDA Forest Service Pacific Southwest Research Station.

Falanruw, M. V. C. 1990b. Traditional Adaptation to Natural Processes of Erosion and Sedimentation on Yap Island. *Proceedings, International Symposium on Research Needs and Applications to Reduce Erosion and Sedimentation in Tropical Steeplands.* International Association of Hydrological Science Press, Wallingford, UK.

Falanruw, M. V. C., J. E. Maka, T. G. Cole and C. D. Whitesell 1990. *Common and Scientific Names of Trees and Shrubs of Mariana, Caroline, and Marshall Islands,* Resource Bulletin, PSW-26. Berkeley, CA: USDA Forest Service Pacific Southwest Research Station.

Falanruw, M. V. C. 1991. Culture and resource management: Factors affecting forests, In E. Conrad and L. Newell (eds.). *Proceedings of a Symposium on Tropical Forestry for People in the Pacific, XVII Pacific Science Congress,* pp. 31–36, General Technical Report, PSW-129. Berkeley, CA: USDA Forest Service.

Falanruw, M. V. C. 1992. Taro growing on Yap. *Conference on Sustainable Taro Culture for the Pacific, September 24–25, 1992,* Honolulu, Hawaii.

Falanruw, M. V. C. 1994. Food production and ecosystem management on Yap. Isla. *Journal of Micronesian Studies* 2(1): 5–22.

Falanruw, M. V. C. 1995. The Yapese agricultural system. *PhD Dissertation.* University of the South Pacific, Suva, Fiji.

Falanruw, M. V. C. 1996. *The Third Nguchol. Paper Presented at the First Yap State Economic and Social Summit.* Colonia: Federated States of Micronesia.

Falanruw, M. and F. Ruegorong. 1997. Indigenous fallow management on Yap islands. In M. Cairns (ed.). *Indigenous Strategies for Intensification of Shifting Cultivation in Asia-Pacific.* Bogor: ICRAF.

Falanruw, M. V. C. 1999. *Terrestrial Biodiversity of the FSM. A Report Prepared for the FSM National Biodiversity Strategy and Action Plan.* Palikir: FSM National Government.

Falanruw, M. V. C. 2005. Chothowliy yuu Waab, take care of Yap, work of the Yap environmental stewardship consortium. *Presentation at 3rd Micronesian Traditional Leadership Conference,* Colonia, Yap.

Falanruw, M. V. C. and C. Chieng. 2005. Some environmental and community activities in the FSM. *24th Pacific Islands Environment Conference Session 10.* University of Guam, Mangilao, Guam.

Falanruw, M. V. C. and F. Ruegorong. 2010. Indigenous fallow management on Yap island, In M. F. Cairns (ed.). *Voices of the Forest, Integrating Indigenous Knowledge into Sustainable Upland Forestry.* London: Taylor & Francis Press.

Falanruw, M. V. C. *The Yap Almanac Calendar 1980–2014.* Colonia: The Yap Institute of Natural Science.

Falanruw, M. V. C. and F. Ruegorong. 2015a. Dynamics of an island ecosystem: Where to now? In M. F. Cairns (ed.). *Shifting Cultivation and Environmental Change: Indigenous People, Agriculture and Forest Conservation.* London: Taylor & Francis Press.

Falanruw, M. V. C. 2015b. *Trees of Yap: A Field Guide. General Technical Report,* PSW-GTR-249. Hilo: USDA Forest Service Pacific Southwest Research Station.

Federated States of Micronesia (FSM). 2002. *The Federated States of Micronesia national Biodiversity Strategy and Action Plan.* Palikir: FSM Government.

Federated States of Micronesia (FSM). 2010. *State-Wide Assessment and Resource Strategy.* Palikir: FSM Government.

Federated States of Micronesia Division of Statistics. 2012. *Summary Analysis of Key Indicators: From the FSM 2010 Census of Population and Housing.* Palikir: FSM Government.

Fletcher, C. H. and B. M. Richmond. 2010. *Climate Change in the Federated States of Micronesia: Food and Water Security, Climate Risk Management and Adaptive Strategies.* Honolulu: Sea Grant, University of Hawaii.

Fosberg, F. R. and M. V. C. Falanruw. 1974. A new Micronesian Terminalia (Combretaceae). *Phytologia* 28(5): 469–470.

Fosberg, F. R., M. V. C. Falanruw, and M. H. Sachet. 1975. *Vascular Flora of the Northern Mariana Islands.* Smithsonian Contributions to Botany, vol. 22, iii+45pp, Washington, DC: Smithsonian Institution.

Gaan, M., C. Chieng, and M. Falanruw. 2004. *Chothowliy yuu Waab.* Colonia: Yap State Biodiversity Strategy and Action Plan, Yap State Government.

Gootnick, D. 2016. *Issues Associated with Implementation in Palau, Micronesia, and the Marshall Islands.* Washington, DC: Government Accountability Office.

Gotz, S., A. B. Gustavo, C. A. H. da Fonseca, C. Gascon, H. C. Vasconcelos, and A. M. N. Izac (eds.). 2004. *Agroforestry and Biodiversity Conservation in Tropical Landscapes.* Washington, DC: Island Press.

Hezel, F. 2009. High water in low atolls. *Micronesian Counselor.* http://www.micsem. org/ checked February 06, 2019.

Hunt, E., N. Kidder, and D. M. Schneider. 1949. *The Micronesians of Yap and Their Depopulation. Report of the Peabody Museum Expedition to the Yap Islands (1947–1948).* Cambridge, MA: Harvard University.

International Panel on Climate Change (IPCC). 2007. *21st Century Global Changes, Topic 3, Section 3.2.1, Fourth Assessment Report, Climate Change Synthesis Report.* Geneva: International Panel on Climate Change.

Johnson, C. G., P. J. Alvis, R. L. Hetzler, U.S. Army, and U.S. Geological Survey. 1960. *Military Geology of Yap Islands, Caroline Islands.* Stanford, CA: Stanford University.

Keener, V. W., J. J. Marra, M. L. Finucane, D. Spooner, and M. H. Smith. 2012. *Climate Change and Pacific Islands: Indicators and Impacts: Executive Summary of the 2012 Pacific Islands Regional Climate Assessment (PIRCA).* Geneva: Island Press.

Keldermans, B., D. Otobed, and M. Falanruw. 1994. The impact of the war and last 50 years and Palau's natural environment. *An International Conference on the War in Palau: 50 Years of Change, Bai ra medal,* Koror, Palau.

Lubchenco, J., A. M. Olson, L. B. Brubaker, S. R. Carpenter, M. M. Holland, S. P. Hubbell, S. A. Levin, J. A. MacMahon, P. A. Matson, J. M. Melillo, H. A. Mooney, C. H. Peterson, H. R. Pulliam, L. A. Real, P. J. Regal, and P. G. Riser. 1991. The sustainable biosphere initiative: An ecological research agenda. *The Ecological Society of America* 72(2): 371–412.

Mace, M. J. 1999. *National Legislation Relevant to Biodiversity within the Federated States of Micronesia.* Palikir: FSM Government.

MacLean, C. D., T. G. Cole, C. D. Whitesell, M. V. C. Falanruw, A. H. Ambacher. 1986. *Vegetation Survey of Pohnpei, Federated States of Micronesia,* Res. Bull. PSW-18. Berkeley, CA: USDA Forest Service.

McClure, J. 2018a. Chinese target Yap fish with some local help. *Pacific Island Times.* www.pacificislandtimes.com/ checked February 06, 2019.

McClure, J. 2018b. *Traditional Yap Chief axed for fishing deals with outsiders*. Pohnpei: Kaselehlie Press.

McNeely, J. A. and S. J. Scherr. 2002. *Ecoagriculture, Strategy to Feed the World and Save wild Biodiversity*. Washington, DC: Island Press.

Meehl, G. A., W. M. Washington, W. D. Collins, J. M. Arblaster, A. Hu, L. B. Buja, W. G. Strand, and H. Teng. 2007. How much more global warming and sea level rise? *Science* 307: 1766–1769.

Morse, J. E., M. Falanruw, C. D. Whitesell, and J. Baldwin. 1987. *Fruit Bats (Megachiroptera) of the World: A Bibliography*. Honolulu: USDA Forest Service Pacific Southwest Research Station.

National Research Council. 1989. *Alternative Agriculture*. Washington, DC: National Academy Press.

Nunn, P. D. 2013. The end of the Pacific? Effects of sea level rise on Pacific Island livelihoods. *Singapore Journal of Tropical Geography* 34(2): 143–171.

Nurse, L. A., R. F. McLean, J. Agard, L. P. Briguglio, V. Duvat-Magnan, N. Pelesikoti, E. Tompkins, and E. Webb. 2014. Small islands, In V. R. Barros, C. B. Field, M. D. Dokken, K. J. Mach, T. E. Bilir, M. Chatterjee, K. L. Ebi, Y. O. Estrada, R. C. Genova, B. Girma, E. S. Kissel, A. N. Levy, S. MacCracken, P. R. Mastrandrea, and L. L. White (eds.). *Climate Change 2014: Impacts, Adaptation, and Vulnerability. Part B: Regional Aspects. Contribution of Working Group II to the Fifth Assessment Report of the Intergovernmental Panel on Climate Change*. Cambridge, UK: Cambridge University Press.

Pacific Islands Climate Education Partnership (PICEP). 2014. *Climate Change in the Federated States of Micronesia*. Honolulu: Pacific Resources for Education and Learning.

Perkins, R. M. and S. M. Krause. 2018. Adapting to climate change impacts in Yap State, Federated States of Micronesia: The importance of environmental conditions and intangible cultural heritage. *Island Studies Journal* 13(1): 65–78.

Pratt, H. D., M. V. C. Falanruw, and M. T. Etpison. 2008. Noteworthy bird observations in Micronesia, including five new records for Micronesia. *Western Birds* 41: 90–101.

Scherr, S. J. and J. A. McNeely. 2007. *Farming with Nature. The Science and Practice of Ecoagriculture*. Washington, DC: Island Press.

Storlazzi, C. D., P. L. E. Edwin, and P. Berkowitz. 2015. Many atolls may be uninhabitable within decades due to climate change. *Scientific Reports* 5: 14546; DOI:10.1038/srep14546.

Terry, J. P. and A. C. Falkland. 2009. Responses of atoll freshwater lenses to storm-surge overwash in the Northern Cook Islands. *Hydrology Journal* 18(3): 749–759.

Tetens, A. 1958. *Among the Savages of the South Seas*. London: Oxford Press.

The Nature Conservancy. 2006. Declaration of commitment: 'The Micronesia Challenge.' TNC Asia-Pacific/Micronesia Program. www.micronesiachallenge.org checked September 15, 2018.

Underwood, J. H. 1969. Preliminary investigations of demographic features and ecological variables of a Micronesian island population. *Micronesia* 5(1): 1–24.

United Nations. 1994. *Report of the Global Conference on the Sustainable Development of Small Island Developing States*. Bridgetown: United Nations.

United Nations Environment Programme (International Assessment of Agricultural Knowledge Science and Technology for Development). 2009. *Agriculture at a Crossroads*. Washington, DC: Island Press.

United States Geological Survey (USGS). 1983. Yap Islands (Waqab). Federated States of Micronesia. Aerial photographs taken in 1969, field checked in 1980.

U.S. National Research Council. 2010. *Sea Level Rise and the Coastal Environment*. Washington, DC: Advancing the Science of Climate Change, National Research Council of the National Academies, The National Academies Press.

U.S. Department of Interior. 2017. *Fiscal Year 2016 Annual Report*. Washington, DC: US Government.

Useem, J. 1946. *Economic and Human Resources, Yap, Palau, West Carolines, United States*. Honolulu: United States Commercial Company Economic Survey.

Volkens, G. 1901. Uber die Karolinen-Insel Yap. *Vorhandlungen der Gosoll-schaft fur Erkundo* 36: 62–76.

Werner, A. D., H. K. Sharp, S. C. Galvis, V. E. A. Post, D. Jacovovic, A. Bosserelle, P. Sinclair. 2017. Hydrology and management of freshwater lenses on atoll Islands: Review of current knowledge and research challenges. *Journal of Hydrology* 551: 819–844.

Whitesell, C., C. D. MacLean, M. V. C. Falanruw, T. G. Cole, and A. H. Ambacher. 1986. *Vegetation Survey of Kosrae, Federated States of Micronesia*, Resource Bull. PSW-17. Berkeley, CA: USDA Forest Service.

Wiles, G. J., J. Engbring, and M. V. C. Falanruw. 1991. Population status and natural history of Pteropus mariannus in Ulithi Atoll, Caroline Islands. *Pacific Science* 45(10): 76–84.

Wilken, C. G. 1987. *Good Farmers: Traditional Agricultural Resource Management in Mexico and Central America*. Berkeley, CA: University of California Press.

Wynn, A., R. Reynolds, D. Buden, M. Falanruw, and B. Lynch. 2012. The unexpected discovery of blind snakes (Serpentes: Typhlopidae) in Micronesia: Two new species of Rhamphotyphlops from the Caroline Islands. *Zootaxa* 3172: 39–54.

Yap State Statistics Office. 2011. *Yap State Statistical Yearbook*. Colonia: Yap State Government, Federated States of Micronesia.

Yap State. 2014. *Criteria for Protected Areas under the Micronesia Challenge*, Colonia: Yap State Government, Federated States of Micronesia.

Chapter 5

Abdul, M., H. Ismail, H. Hashim, and J. Johari. 2009. Consumer decision making process in shopping for halal food in Malaysia. *China-USA Business Review* 8(9): 40–47.

Ahmed, A. (2008). Marketing of halal meat in the United Kingdom: Supermarkets versus local shops. *British Food Journal* 110(7): 655–670.

Al-Jowder, O., E. K. Kemsley, and R. H. Wilson. 1997. Mid-infrared spectroscopy and authenticity problems in selected meats: A feasibility study. *Journal of Food Chemistry* 59: 195–201.

Al-Qaradawi, Y. 1984. *The Lawful and the Prohibited in Islam*. Beirut: The Holy Quran Publishing House.

Anderson, E. W., C. Fornell, and D. R. Lehmann. 1994. Customer satisfaction, market share, and profitability: Findings from Sweden. *The Journal of Marketing* 58: 53–66.

Baby, R. E., M. Cabezas, and E. N. Walsoe de Reca. 2000. Electronic nose: A useful tool for monitoring environmental contamination. *Sensors and Actuators B: Chemistry* 69: 214–218.

Briandet, R., E. K. Kemesly, and R. H. Wilson. 1996. Discrimination of Arabica and Robusta in instant coffees by Fourier Transform Infrared Spectroscopy and chemometrics for the authentication of fruit purees. *Journal of Agricultural Food Chemistry* 44: 170–174.

Capone, S., M. Epifani, F. Quaranta, P. Siciliano, A. Taurino, and L. Vasanelli. 2001. Monitoring of rancidity of milk by means of an electronic nose and a dynamic PCA analysis. *Sensors and Actuators B: Chemistry* 78: 174–179.

Che Man, Y. B. and G. Setiowaty. 1999a. Application of FTIR transmission spectroscopy in determining FFA content of palm olein. *Food Chemistry* 66(1): 109–114.

Che Man, Y. B. and G. Setiowaty. 1999b. Determination of anisidine value of thermally oxidized palm oil by FTIR spectroscopy. *Journal of American Oil Chemists' Society* 76: 243–247.

Che Man, Y. B. and M. E. S. Mirghani. 2000. Detection of lard mixed with body fats of chicken, lamb, and cow by fourier transform infrared spectroscopy. *Journal of American Oil Chemists' Society* 78(7): 753–761.

Coni, E., M. D. Pasquale, P. Coppolelli, and A. Bocca. 1994. Detection of animal fats in butter by differential scanning calorimetry: A pilot study. *Journal of American Oil Chemists' Society* 71: 807–810.

Defernez, M., E. K. Kemesly, and R. H. Wilson. 1995. The use of FTIR and chemometrics for the authentication of fruit purees. *Journal of Agricultural and Food Chemistry* 43: 109–113.

Defernez, M. and R. H. Wilson. 1995. Mid-infrared spectroscopy and chemometrics for determining the type of fruit used in jam. *Journal of Agriculture and Food Chemistry* 67: 461–467.

DeMan, J. M. 1999. Functionality requirements of fats and oils for food applications. *Paper Presented at MOSTA Tech-In, Recent Advances in the Sciences of Oils and Fats; Malaysian Oil Scientists' and Technologists' Association.* Petaling Jaya, Selangor DE, Malaysia.

Dindyal, S. and S. Dindyal. 2003. How personal factors, including culture and ethnicity, affect the choices and selection of food we make. *Internet Journal of World Medicine* 1(2): 27–33.

Folkes, D. J. and L. Crane. 1988. Determination of carbohydrates, In R. Macrae (ed.). *HPLC in Food Analysis*, 2nd edition. San Diego, CA: Academic Press.

Haryati, T. 1999. Development and application of differential scanning calorimetric technique for physical and chemical analysis of palm oil. *Doctor of Philosophy Thesis.* Faculty of Food Science and Biotechnology, Universiti Putra Malaysia, Serdang, Selangor DE, Malaysia.

Hermann, S. 1997. Hysteresis in marketing-a new phenomenon? *MIT Sloan Management Review* 38(3): 39.

Lambelet, P. 1983. Detection of pig and buffalo body fat in cow and buffalo ghees by differential scanning calorimetry. *Journal of American Oil Chemists' Society* 60: 1005–1008.

Lambelet, P., O. P. Singhal, and N. C. Ganguli. 1980. Detection of goat body fat in ghee by differential thermal analysis. *Journal of American Oil Chemists' Society* 57: 364–366.

Lambert, W. E., H. J. Nelis, M. G. M. de Ruyter, and A. P. Leenheer. 1985. *Journal of Chromatographic Science* 30: 1–72.

Macrae, R. 1988. HPLC, In R. Macrae (ed.). *Food Analysis*, 2nd edition. San Diego, CA: Academic Press.

Marikkar, J. M. N., O. M. Lai, H. M. Ghazali, and Y. B. Che Man. 2001. Detection of lard and randomization lard as adulterants in refined-bleached-deodorized palm oil by differential scanning calorimetry. *Journal of American Oil Chemists' Society* 78: 1113–1119.

Marikkar, J. M. N., O. M. Lai, H. M. Ghazali, and Y. B. Che Man. 2002. Compositional and thermal analysis of RBD palm oil adulterated with lipase-catalyzed interesterified lard. *Food Chemistry* 76: 249–258.

Regnier, F. E. 1983. HPLC of proteins, peptides, and polynucleotides. *Analytical Chemistry* 55(13): 1298A–1306A.

Santos, C. P. D. and D. V. D. H. Fernandes. 2008. Antecedents and consequences of consumer trust in the context of service recovery. *Brazilian Administration Review* 5(3): 225–244.

Shen, N., S. Moizuddin, L. Wilson, S. Duvick, P. White, and L. Pollak. 2001. Relationship of electronic nose analysis and sensory evaluation of vegetable oils during storage. *Journal of American Oil Chemists' Society* 78: 937–940.

Staples, E. J. 2001. The Znose™: The first quantitative electronic nose with olfactory images and virtual sensor arrays. *Paper #9671 Presented at The Institute of Food Technologists*, New Orleans, LA, USA.

Wang, D. Q. and E. Kolbe. 1991. Thermal properties of surimi analyzed using DSC. *Journal of Food Science* 56: 302–308.

Wilson, R. H. and B. J. Goodfellow. 1994. Mid-infrared spectroscopy, In R. H. Wilson (ed.). *Spectroscopic Techniques for Food Analysis*. New York: CRC Press.

Yoke Wah, L., E. K. Kemsley, and R. H. Wilson. 1994. The potential of fourier transform infrared spectroscopy for the authentication of edible oils. *Journal of Agriculture and Food Chemistry* 42: 1154–1159.

Chapter 6

AccountAbility. 2011. AA1000 Stakeholder engagement standard 2011. www.accountability.org checked September 15, 2018.

Aldrich, J. M., W. R. Ostlie, and T. M. Faust. 1997. *Great Plains Landscapes of Biological Significance*. Arlington, VA: The Nature Conservancy.

Bruins, R. J., S. E. Franson, W. E. Foster, F. B. Daniel, and P. B. Woodbury. 2009. *A Methodology for the Preliminary Scoping of Future Changes in Ecosystem Services, With an Illustration from the Future Midwestern landscapes study*, EPA/600/R-09/134. Washington, DC: U.S. Environmental Protection Agency.

Bush, G. W. 2006. State of the union address. https://georgewbush-whitehouse.archives.gov/stateoftheunion/2006/ checked September 15, 2018.

Chen, M., P. M. Smith, and M. P. Wolcott. 2016. U.S. biofuels industry: A critical review of opportunities and challenges. *BioProducts Business* 1: 42–59.

Clark, J. S. 1996. The great plains partnership, In F. B. Samson and F. L. Knopf (eds.). *Prairie Conservation*. Washington, DC: Island Press.

Cooperrider, D. L., D. K. Whitney, and J. M. Stavros. 2003. *Appreciative Inquiry Handbook: The First in a Series of AI Workbooks for Leaders of Change.* San Francisco, CA: Berrett-Koehler Publishers.

Cooter, E. J., R. Dodder, J. Bash, A. Elobeid, L. Ran, V. Benson, and D. Yang. 2017. Exploring a United States maize cellulose biofuel scenario an integrated energy and agricultural markets solution approach. *Annals of Agricultural & Crop Sciences* 2: 1031–1042.

Costanza, R., R. d'Arge, R. de Groot, S. Farber, M. Grasso, B. Hannon, K. Limburg, S. Naeem, R. V. O'Neill, J. Paruelo, R. G. Raskin, P. Sutton, and M. van den Belt. 1997. The value of the world's ecosystem services and natural capital. *Nature* 387: 253–260.

Coyle, W. T. 2007. *The Future of Biofuels: A Global Perspective.* Washington, DC: U.S. Department of Agriculture, Economic Research Service. www.ers.usda. gov/amber-waves/ 2007/november/the-future-of-biofuels-a-global-perspective/ checked September 15, 2018.

Cruse, R. 2007. Biofuel feedstock landscape coverage and ecoservice implications for various biofuel industry scenarios. Application for Federal Assistance (SF 424) submitted to USEPA Region 7, June 7, 2007. Iowa State University, Ames, IA, USA.

Dale, V. D. and S. Polasky. 2007. Measures of the effects of agriculture practices on ecosystem services. *Ecological Economics* 64: 286–296.

DeCicco, J. M., D. Y. Liu, J. Heo, R. Krishnan, A. Kurthen, and L. Wang. 2016. Carbon balance effects of U.S. biofuel production and use. *Climate Change* 138: 667–680.

Devadoss, S., P. Westhoff, and M. Helmar, E. Grundmeier, and K. Skold. 1989. *The FAPRI Modeling System at CARD: A Documentation Summary.* Ames, IA: Iowa State University Center for Agricultural and Rural Development.

Dietz, T., E. Ostrom, and P. C. Stern. 2009. The struggle to govern the commons. *Science* 302: 1907–1912.

Dodder, R. S., A. E. Elobeid, T. L. Johnson, P. O. Kaplan, and L. A. Kurkalova. 2011. *Environmental Impacts of Emerging Biomass Feedstock Markets: Energy, Agriculture, and the Farmer.* Working paper 11-WP 526. Ames, IA: Iowa State University Center for Agriculture and Rural Development.

Dodder, R. S., P. O. Kaplan, A. Elobeid, S. Tokgoz, S. Secchi, and L. A. Kurkalova. 2015. Impact of energy prices and cellulosic biomass supply on agriculture, energy, and the environment: An integrated modeling approach. *Energy Economics* 51: 77–87.

Duffield, J. A. and K. Collins. 2006. Evolution of renewable energy policy. *Choices* 1: 9–11.

Goldman, R. L., B. H. Thompson, and G. C. Daily. 2007. Institutional incentives for managing the landscape: Inducing cooperation for the production of ecosystem services. *Ecological Economics* 64: 333–343.

Gray, G. 2007. Comments during U.S. Environmental Protection Agency. *National Council of Environmental Policy and Technology Energy and Environment Workgroup April 26–27, 2007 Meeting*, Washington, DC, USA.

Great Plains Partnership and Western Governors' Association. 1996. Vision for the future. *Meeting Materials for the Great Plains Partnership Council June 21–22, 1996 Meeting*, Omaha, Nebraska.

Grindstaff, S. G. and Groskinsky, B. L. 2015. Achieving meaningful stakeholder dialogues: The U.S. Environmental Protection Agency's use of new engagement mechanisms for understanding stakeholder perception ad interaction using enhanced place-based appreciative methods. *Alternative Dispute Resolution Committee Newsletter July 11, 2015,* Section of Environment, Energy, and Resources, American Bar Association, Chicago, IL, USA.

Grindstaff, S. G. and Groskinsky, B. L. 2016. Achieving meaningful stakeholder dialogues: Environmental research planning using new enhanced place-based methods of engagement. *Alternative Dispute Resolution Committee Newsletter January 12, 2016,* Section of Environment, Energy, and Resources, American Bar Association, Chicago, IL, USA.

Groskinsky, B. 2007. U.S. EPA Region 7 comments on biofuels. *Presentation at U.S. Environmental Protection Agency National Council of Environmental Policy and Technology Energy and Environment Workgroup August 8–10, 2007 Meeting,* Boston, MA, USA.

Hammes, D. and D. Willis. 2005. Black gold: The end of Bretton Woods and the oil price shocks of the 1970s. *Independent Review* 4: 501–511.

Johnson, T. L., J. F. DeCarolis, C. L. Shay, D. H. Loughlin, C. L. Gage, and S. Vijay. 2006. *MARKAL Scenario Analysis of Technology Options of the Electric Sector: The Impact on Air Quality,* EPA-600/R-06/114. Washington DC: U.S. Environmental Protection Agency.

Kansas State University Center for Sustainable Energy. 2007. *About Us.* Manhattan, KS: Kansas State University.

Linthurst, R. 2007. Overview of ecological research program. *Presentation at U.S. Environmental Protection Agency Science Advisory Board July 2007 Meeting,* Washington, DC, USA.

Liska, A. J. and R. K. Perrin. 2009. Indirect land use emissions in the life cycle of biofuels: Regulations vs science. *Biofuels, Bioproducts and Biorefining* 3: 318–328.

McManus, M. C. and C. M. Taylor. 2015. The changing nature of life cycle assessment. *Biomass and Bioenergy* 82: 13–26.

Mehaffey, M., E. Smith, and R. Van Remortel. 2012. Midwest U.S. landscape change to 2020 driven by biofuel mandates. *Ecological Applications* 22: 8–19.

Mehaffey, M., R. Van Remortel, E. Smith, and R. Bruins. 2011. Developing a dataset to assess ecosystem services in the Midwest, United States. *International Journal of Geographic Information Services* 25: 681–695.

Merrill, K. R. 2007. *The Oil Crisis of 1973–1974: A Brief History with Documents.* Boston, MA: Bedford/St. Martin's.

Millennium Ecosystem Assessment. 2005. Statement of the MA Board—Living beyond our means: Natural assets and human well-being. www.millenniumassessment.org/en/BoardStatement.html checked September 15, 2018.

National Research Council of the National Academies. 2008. *Transitioning to Sustainability through Research and Development of Ecosystem Services and Biofuels: Workshop Summary.* Washington, DC: The National Academies Press.

Nelson, R. G. 2002. Resource assessment and removal analysis for corn stover and wheat straw in the Eastern and Midwestern United States—rainfall and wind-induced soil erosion methodology. *Biomass & Bioenergy* 22: 349–363.

Nelson, R. G. and M. D. Schrock. 2006. Energetic and economic feasibility associated with the production, processing and conversion of beef tallow to a substitute diesel fuel. *Biomass & Bioenergy* 30: 584–591.

Osborn, C. T., F. Llacuna, and M. Linsenbigler. 1992. *The Conservation Reserve Program: Enrollment Statistics for Signup Periods 1–11 and Fiscal Years 1990–92*, SB-843. Washington, DC: U.S. Department of Agriculture, Economic Research Service.

Ostlie, W. R., R. E. Schneider, J. M. Aldrich, T. M. Faust, R. L. B. McKim, and S. J. Chaplin. 1997. *The Status of Biodiversity in the Great Plains*. Arlington, VA: The Nature Conservancy.

Pendell, D. L., J. R. Williams, S. B. Boyles, C. W. Rice, and R. G. Nelson. 2007. Soil carbon sequestration strategies with alternative tillage and nitrogen sources under risk. *Applied Economic Perspectives and Policy* 29: 245–268.

Reilly, W. K. 1996. Across the barricades, In H. L. Diamond and P. F. Noonan (eds.). *Land Use in America—The Report of the Sustainable Use of Land Project*. Washington, DC: Island Press.

Riley, C., R. Wooley, and D. Sandor. 2007. Implementing systems engineering in the U.S. Department of Energy Office of the Biomass Program. *2007 International Conference on System of Systems Engineering*, SOSE in Service of Energy and Security, San Antonio, TX, USA.

Rogovska, N., D. Laird, R. Cruse, P. Fleming, T. Parkin, and D. Meek. 2011. Impact of biochar on manure carbon stabilization and greenhouse gas emissions. *Soil Science Society of America Journal* 75: 871–879.

Rustigian, H. L., M. V. Santelmann, and N. H. Schumaker. 2003. Assessing the potential impacts of alternative landscape designs on amphibian population dynamics. *Landscape Ecology* 18: 65–81.

Sampson, F. B. and F. L. Knoff (eds.). 1996. *Prairie Conservation—Preserving North America's Most Endangered Ecosystem*. Washington, DC: Island Press.

Santelmann, M., K. Freemark, J. Sifneos, and D. White. 2006. Assessing effects of alternative agricultural practices on wildlife habitat in Iowa, U.S.A. *Agriculture, Ecosystems and Environment* 113: 243–253.

Santelmann, M. V., D. White, K. Freemark, J. I. Nassauer, J. M. Eilers, K. B. Vaché, B. J. Danielson, R. C. Corry, M. E. Clark, S. Polasky, R. M. Cruse, J. Sifneos, H. Rustigian, C. Coiner, J. Wu, and D. Debinski. 2004. Assessing alternative futures for agriculture in Iowa, U.S.A. *Landscape Ecology* 19: 357–374.

Savage, C. 2011. *Prairie—A Natural History*. Vancouver, BC. Greystone Books.

Schneider, R. E., D. Faber-Langendoen, R. C. Crawford, and A. S. Weakley. 1997. *Great Plains Vegetation Classification*. Arlington, VA: The Nature Conservancy.

Scholtz, R., J. A. Polo, S. D. Fuhlendorf, and G. D. Duckworth. 2017. Land cover dynamics influence distribution of breeding birds in the Great Plains, USA. *Biological Conservation* 209: 323–331.

Seavey, C. A. 2003. The American public library during the great depression. *Library Review* 8: 373–378.

Secchi, S., J. Tyndall, L. A. Schulte, and H. Asbjornsen. 2008. High crop prices and conservation—raising the stakes. *Journal of Soil and Water Conservation* 3: 68A–73A.

Tenenbaum, D. J. 2008. Food vs. fuel: Diversion of crops could cause more hunger. *Environmental Health Perspectives* 116: A254–A257.

The Economist. 2009. The bigger picture. www.economist.com/finance-and-economics/2009/10/12/the-bigger-picture checked September 2018.

The Forrester Group. 2007. *Future Midwestern Landscape Study Region 7 Stakeholder Forum—November 7, 2007*. Chesterfield, MO: The Forrester Group.

The Harwood Group, Great Plains Partnership and Western Governors' Association. 1996. *A Way of Life: Great Plains citizens Talk about Ecosystems. A Great Plains Partnership Report.* Bethesda, MD: The Harwood Group.

The White House National Economic Council. 2006. Advanced energy initiative. https://georgewbush-whitehouse.archives.gov/ceq/advanced-energy.html checked September 15, 2018.

Trostle, R. 2008. *Global Agricultural Supply and Demand: Factors Contributing to the Recent Increase in Food Commodity Prices. Economic Research Service Report.* USDA/WRS-0801. Washington, DC: USDA.

U.S. Congress. 1978. Energy tax act of 1978. www.congress.gov/bill/95th-congress/house-bill/5263 checked September 15, 2018.

U.S. Congress. 1980. Energy security act of 1980. www.congress.gov/bill/96th-congress/senate-bill/932 checked September 15, 2018.

U.S. Congress. 1987. National energy conservation policy act of 1987. www.congress.gov/bill/95th-congress/house-bill/5037 checked September 15, 2018.

U.S. Congress. 1999. Biomass research and development (BR&D) act of 1999–2000. www.congress.gov/bill/106th-congress/house-bill/2819 checked September 15, 2018.

U.S. Congress. 2005. Energy policy act of 2005. www.congress.gov/bill/109th-congress/house-bill/6 checked September 15, 2018.

U.S. Congressional Budget Office. 2009. The impact of ethanol use on food prices and greenhouse-gas emissions. www.cbo.gov/sites/default/files/111th-congress-2009-2010/reports/04-08-ethanol.pdf checked September 15, 2018.

U.S. Department of Agriculture. 2007. *Agricultural Projections to 2016.* OCE-2007-1. Washington, DC: USDA. www.ers.usda.gov/webdocs/publications/37775/11618_oce20071_1_.pdf?v=41056 checked September 15, 2018.

U.S. Department of Agriculture. 2013. Conservation reserve program fact sheet. www.fsa.usda.gov/Internet/FSA_File/crpfactsheet0213.pdf checked September 15, 2018.

U.S. Department of Agriculture. 2016. USDA invests $1.7 billion to project sensitive agricultural lands through Conservation Reserve Program. Press Release No. 0233.16 issued October 28, 2016.

U.S. Department of Commerce, Economics and Statistics Administration, U.S. Census Bureau. 1984. Census regions and divisions of the United States. www2.census.gov/geo/pdfs/maps-data/maps/reference/us_regdiv.pdf checked September 15, 2018.

U.S. Department of Energy. 2005. *Biomass as Feedstock for a Bioenergy and Bioproducts Industry: The Technical Feasibility of a Billion-Ton Annual Supply.* ORNL/TM-/66. Oak Ridge, TN: Oak Ridge National Laboratory.

U.S. Department of Energy. 2011. *U.S. Billion-Ton Update: Biomass Supply for a Bioenergy and Bioproducts Industry*, ORNL/TM-2011-224. Oak Ridge, TN: Oak Ridge National Laboratory.

U.S. Department of Energy. 2016. *2016 Billion-Ton Report: Advancing Domestic Resources for a Thriving Bioeconomy, Volume 1: Economic Availability of Feedstocks and Maps and Data on the Bioenergy Knowledge Discovery Framework, Volume 2: Environmental Availability of Select Scenarios*, ORNL/TM-2016/160. Oak Ridge, TN: Oak Ridge National Laboratory.

U.S. Department of Energy Office of Energy Efficiency & Renewable Energy. 2016. 2016 billion-ton report. www.energy.gov/eere/bioenergy/ 2016-billion-ton-report checked September 15, 2018.

U.S. Environmental Protection Agency. 1970. Summary of the clean air act. www.epa.gov/laws-regulations/summary-clean-air-act checked September 15, 2018.

U.S. Environmental Protection Agency. 1986. Toxic Release Inventory (TRI) program. www.epa.gov/toxics-release-inventory-tri-program checked September 15, 2018.

U.S. Environmental Protection Agency. 1990. *Reducing Risk: Setting Priorities and Strategies for Environmental Protection*, SAB-EC-90-21. Washington, DC: U.S. Environmental Protection Agency, Science Advisory Board.

U.S. Environmental Protection Agency. 2006. *Life Cycle Assessment: Principles and Practice*, EPA/600/R-06/0600. Washington, DC: U.S. Environmental Protection Agency.

U.S. Environmental Protection Agency Ecological Research Program. 2007. Ecological research program future Midwestern landscapes study focuses on ecosystem services. https://nepis.epa.gov/Exe/ZyPDF.cgi/P10060Z2. PDF?Dockey=P10060Z2.PDF checked September 15, 2018.

U.S. Environmental Protection Agency National Advisory Council for Environmental Policy and Technology Energy and Environment Workgroup. 2007. Charge to the national advisory council for environmental policy and technology energy and environment workgroup. www.epa.gov/faca/nacept checked September 15, 2018.

U.S. Environmental Protection Agency Region 7. 1999. *Remember the Past Protect the Future*, EPA-903-R-99-005. Kansas City, KN: U.S. Environmental Protection Agency Region 7.

U.S. Environmental Protection Agency Region 7. 2007. *Environmental Laws Applicable to Construction and Operation of Ethanol Plants*, EPA-907-B-07-001. Lenexa, KS: U.S. Environmental Protection Agency Region 7. https://archive.epa.gov/ncea/ biofuels/web/pdf/ethanol_plants_manual.pdf checked September 15, 2018.

U.S. Environmental Protection Agency Region 7. 2008. *Environmental Laws Applicable to Construction and Operation of Biodiesel Production Facilities*, EPA-901-B-001. Lenexa, KS: U.S. Environmental Protection Agency Region 7. https://archive. epa.gov/ncea/biofuels/web/pdf/biodiesel_manual.pdf checked September 15, 2018.

U.S. Environmental Protection Agency Renewable Fuel Standard Program. 2017. Proposed renewable fuel standards for 2017, and the biomass-based diesel volume for 2018. www.epa.gov/renewable-fuel-standard-program/proposed-renewable-fuel-standards-2017-and-biomass-based-diesel checked September 15, 2018.

U.S. Government Publishing Office. 2001. National Energy Policy—Report of the National Energy Policy Development Group. https://wtrg.com/EnergyReport/ National-Energy-Policy.pdf checked September 15, 2018.

Van Sickle, J., J. Baker, A. Herlihy, P. Bayley, S. Gregory, P. Haggerty, L. Ashkenas, and J. Li. 2004. Projecting the biological condition of streams under alternative scenarios of human land use. *Ecological Applications* 14: 368–380.

Williams, J. R., R. G. Nelson, M. M. Claassen, and C. W. Rice. 2004. Carbon sequestration in soil with consideration of CO_2 emissions from production inputs: An economic analysis. *Environmental Management* 33: S264–S273.

Wright, C. K. and M. C. Wimberly. 2013. Recent land use change in the Western Corn Belt threatens grasslands and wetlands. *Proceedings of the National Academy of Sciences* 110: 4134–4139.

Zhang, W., E. A. Yu, S. Rozelle, J. Yang, and S. Msangi. 2013. The impact of biofuel growth on agriculture: Why is the range of estimates so wide? *Food Policy* 38: 227–239.

Chapter 7

Abdullah, K., A. M. Said, and D. Omar. 2014. Community-based conservation in managing mangrove rehabilitation in Perak and Selangor. *Procedia-Social and Behavioral Sciences* 153: 121–131.

Alongi, D. M. 1996. The dynamics of benthic nutrient pools and fluxes in tropical mangrove forests. *Journal of Marine Research* 54: 123–148.

Alongi, D. M. 2014. Carbon cycling and storage in mangrove forests. *Annual Review of Marine Science* 6: 195–219.

Balick, M. J. (ed.). 2009. *Ethnobotany of Pohnpei: Plants, People, and Island Culture.* New York: New York Botanical Garden/University of Hawai'i Press.

Bandaranayake, W. 1998. Traditional and medicinal uses of mangroves. *Mangroves and Salt Marshes* 2: 133–148.

Bann, C. 1997. *An Economic Analysis of Alternative Mangrove Management Strategies in Koh Kong Province, Cambodia.* Singapore: Economy and Environment Program for Southeast Asia Research Report Series.

Barbier, E. 2006. *Mangrove Dependency and the Livelihoods of Coastal Communities in Thailand. Environment and Livelihoods in Tropical Coastal Zones: Managing Agriculture-Fishery-Aquaculture Conflicts.* Wallingford, Oxfordshire: Center for Agriculture and Biosciences International.

Baumgartner, U., S. Kell, and T. H. Nguyen. 2016. Arbitrary mangrove-to-water ratios imposed on shrimp farmers in Vietnam contradict with the aims of sustainable forest management. *SpringerPlus* 5: 438.

Binh, C. T., M. J. Phillips, and H. Demaine. 1997. Integrated shrimp-mangrove farming systems in the Mekong delta of Vietnam. *Aquaculture Research* 28: 599–610.

Brown, B., R. Fadillah, Y. Nurdin, I. Soulsby, and R. Ahmad. 2014. Case study: Community Based Ecological Mangrove Rehabilitation (CBEMR) in Indonesia. From small (12–33 ha) to medium scales (400 ha) with pathways for adoption at larger scales (>5000 ha). *Surveys and Perspectives Integrating Environment and Society* 7(2): 2014.

Dahdouh-Guebas, F., S. Hettiarachchi, D. L. Seen, O. Batelaan, S. Sooriyarahchi, L. P. Jayatissa, and N. Koedam. 2005. Transitions in ancient inland freshwater resource management in Sri Lanka affect biota and human populations in and around coastal lagoons. *Current Biology* 15: 579–586.

Datta, D., R. Chattopadhyay, and P. Guha. 2012. Community based mangrove management: A review on status and sustainability. *Journal of Environmental Management* 107: 84–95.

Donato, D. C., J. B. Kauffman, D. Murdiyarso, S. Kurnianto, M. Stidham, and M. Kanninen. 2011. Mangroves among the most carbon-rich forests in the tropics. *Nature Geoscience* 4: 293–297.

Duke, N. C. 1992. Mangrove floristics and biogeography, In A. I. Robertson and D. M. Alongi (eds.). *Tropical Mangrove Ecosystems*. Washington, DC: American Geophysical Union.

Duke, N. C., J. O. Meynecke, S. Dittman, A. M. Ellison, K. Anger, U. Berger, S. Cannicci, K. Diele, K. C. Ewel, C. D. Field, N. Koedam, S. Y. Lee, C. Marchand, I. Nordhaus, and F. Dahdouh-Guebas. 2007. A world without mangroves? *Science* 317: 41–42.

FAO. 1994. *Mangrove Forest Management Guidelines*, FAO Forestry Paper No. 117. Rome: Food and Agriculture Organization of the United Nations.

FAO. 2007. *The World's Mangroves: 1980–2005*, FAO Forestry Paper 153. Rome: Food and Agriculture Organization of the United Nations.

Giri, C., J. Long, S. Abbas, R. M. Murali, F. M. Qamer, B. Pengra, and D. Thau. 2015. Distribution and dynamics of mangrove forests of South Asia. *Journal of Environmental Management* 148: 101–111.

Giri, C., E. Ochieng, L. L. Tieszen, Z. Zhu, A. Singh, T. Loveland, J. Masek, and N. Duke. 2011. Status and distribution of mangrove forests of the world using earth observation satellite data. *Global Ecology and Biogeography* 20: 154–159.

Government of Cambodia. 2018. Eco adventures Cambodia. https://koh-kong-cambodia.com/kohkongplaces/eco-adventure-cambodia.html checked September 15, 2018.

Ha, T. T. P., H. van Dijk, R. Bosma, and L. X. Sinh. 2013. Livelihood capabilities and pathways of shrimp farmers in the Mekong Delta, Vietnam. *Aquaculture Economics & Management* 17: 1–30.

Hai, T., P. Duc, V. Son, T. Minh, and N. Phuong. 2015. Innovation in seed production and farming of marine shrimp in Vietnam. *World Aquaculture* 46: 32–37.

Hawkins, S., P. X. To, P. X. Phuong, P. T. Thuy, N. D. Tu, C. V. Cuong, S. Brown, P. Dart, S. Robertson, N. Vu, and R. McNally. 2010. *Roots in the Water: Legal Frameworks for Mangrove Payment for Ecosystem Services (PES) in Vietnam*. Washington, DC: Katoomba Group's Legal Initiative Country Study Series, Forest Trends.

Hong, P. N., N. T. K. Cuc, and V. T. Hien. 2008. *Mangrove Restoration for Climate Change Adaptation and Sustainable Development*. Agricultural Publishing House.

I.U.C.N. 2016. Mangroves for the Future in Cambodia.MFF Cambodia factsheet. I.U.C.N.

Kosal, M. 2004. Biodiversity of Cambodia's wetlands, In M. Torrell, A. Salamanca, and B. Ratner (eds.). *Wetlands Management in Cambodia: Socioeonomic, Ecological and Policy Perspectives*, Technical Report 64. Penang: WorldFish Center.

Lacerda, L. D. and Y. Schaeffer-Novelli. 1999. Mangroves of Latin America: The need for conservation and sustainable utilization, In A. Yáñez-Arancibla and A. L. Lara-Domínquez (eds.). *Ecosistemas de Manglar en América Tropical*. Costa Rica: Instituto de Ecología A.C. México, UICN/ORMA.

López-Angarita, J., C. M. Roberts, A. Tilley, J. P. Hawkins, and R. G. Cooke. 2016. Mangroves and people: Lessons from a history of use and abuse in four Latin American countries. *Forest Ecology and Management* 368: 151–162.

López-Hoffman, L., I. E. Monroe, E. Narváez, M. Martínez-Ramos, and D. D. Ackerly. 2006. Sustainability of mangrove harvesting: How do harvesters' perceptions differ from ecological analysis? *Ecology and Society* 11(2): 14.

Lugo, A. E. 2002. Conserving Latin American and Caribbean mangroves: Issues and challenges. *Madera y Bosques* 8: 5–25.

Lugo, A. E. and S. C. Snedaker. 1974. The ecology of mangroves. *Annual Review of Ecology and Systematics* 5: 39–64.

MacKenzie, R. A. and N. Cormier. 2012. Stand structure influences nekton community composition and provides protection from natural disturbance in Micronesian mangroves. *Hydrobiologia* 685: 155–171.

Marschke, M. 1999. Using local environmental knowledge: A case-study of mangrove resource management practices in Peam Krasaop Wildlife Sanctuary, Cambodia. *Master's Thesis*. Dalhousie University, Halifax, Nova Scotia.

Marschke, M. and K. Nong. 2003. Adaptive co-management: Lessons from coastal Cambodia. *Canadian Journal of Development Studies/Revue* 24: 369–383.

Mastaller, M. 1999. *Environmental Management of the Coastal Zone of Cambodia: Assessment of Sustainable Livelihood Alternatives to Mangrove Exploitation.* Copenhagen: DANIDA/Ministry of Environment: Kampsak International.

McIvor, A. L., T. Spencer, I. Möller, and M. R. Spalding. 2012. *Storm Surge Reduction by Mangroves. Natural Coastal Protection Series: Report 2 Cambridge Coastal Research Unit Working Paper 41.* Cambridge, UK: The Nature Conservancy and Wetlands International.

MERC. 2016. Mangroves-based aquaculture in Quang Ninh Province. Final report-IUCN VN/378/2015.

Murdiyarso, D., J. Purbopuspito, J. B. Kauffman, M. W. Warren, S. D. Sasmito, D. C. Donato, S. Manuri, H. Krisnawati, S. Taberima, and S. Kurnianto. 2015. The potential of Indonesian mangrove forests for global climate change mitigation. *Nature Climate Change* 5: 1089–1092.

M.R.C. 2018. Report on the mangrove forest environment of Cambodia www.mekonginfo.org/assets/midocs/0001627-environment-mangrove-forest.pdf checked September 15, 2018.

Nagelkerken, I., S. J. M. Blaber, S. Bouillon, P. Green, M. Haywood, L. G. Kirton, J. O. Meynecke, J. Pawlik, H. M. Penrose, A. Sasekumar, and P. J. Somerfield. 2008. The habitat function of mangroves for terrestrial and marine fauna: A review. *Aquatic Botany* 89: 155–185.

Nasuchon, N. and A. Charles. 2010. Community involvement in fisheries management: Experiences in the Gulf of Thailand countries. *Marine Policy* 34: 163–169.

Naylor, R. and M. Drew. 1998. Valuing mangrove resources in Kosrae, Micronesia. *Environment and Development Economics* 3: 471–490.

Paul, N. C. 1998. Mangrove area encroachment in Cambodia: Problems and Findings report. *Crossing Boundaries, the Seventh Biennial Conference of the International Association for the Study of Common Property,* Vancouver, BC, Canada.

Prahl, H. 1989. *Manglares de Colombia.* Bogotá: Villegas Editores.

Primavera, J. H. 1998. Mangroves as nurseries: Shrimp populations in mangrove and non-mangrove habitats. *Estuarine, Coastal and Shelf Science* 46: 457–464.

Que, N. D. and V. V. Dai. 2015. *Establishment of Coastal Protection Mangrove Forests: Current Status and Solutions.* Hanoi: Vietnam National Agriculture Extension Center, The Ministry of Agriculture and Rural Development.

Reimer, J. K. and P. Walter. 2013. How do you know it when you see it? Community-based ecotourism in the Cardamom Mountains of southwestern Cambodia. *Tourism Management* 34: 122–132.

Rönnbäck, P. 1999. The ecological basis for economic value of seafood production supported by mangrove ecosystems. *Ecological Economics* 29: 235–252.

Rönnbäck, P., M. Troell, N. Kautsky, and J. H. Primavera. 1999. Distribution pattern of shrimps and fish among Avicennia and Rhizophora microhabitats in the Pagbilao Mangroves, Philippines. *Estuarine, Coastal, and Shelf Science* 48: 223–234.

Rotich, B., E. Mwangi, and S. Lawry. 2016. *Where Land Meets the Sea: A Global Review of the Governance and Tenure Dimensions of Coastal Mangrove Forests.* Washington, DC: United States Agency for International Development.

Spalding, M. 2010. *World Atlas of Mangroves.* New York: Routledge.

Tieng, T., S. Sharma, R. A. MacKenzie, M. Venkattappa, N. K. Sasaki, and A. Collin. In review. Improved mapping of mangrove cover using GoogleEarth Engine: Cambodia as a case example.

Tomlinson, P. B. 1986. *The Botany of Mangroves.* Cambridge, UK: Cambridge University Press.

Torell, M., A. M. Salamanca, and B. D. Ratner. 2004. Wetlands management in Cambodia: Socioeconomic, ecological, and policy perspectives. *WorldFish Center Technical Report 64.*

Tuan, L. X., P. N. Hong, and T. Q. Hoc. 2008. Coastal environment issues and mangrove restoration in Vietnam. *Proceedings of 3rd International Conference in Vietnamese Studies: Natural Resources, Environment and Sustainable Development,* Hanoi, Vietnam, pp. 678–692.

Vance, D. J., M. D. E. Haywood, D. S. Heales, R. A. Kenyon, N. R. Loneragan, and R. C. Pendrey. 1996. How far do prawns and fish move into mangroves? Distribution of juvenile banana prawns Penaeus merguiensis and fish in a tropical mangrove forest in northern Australia. *Marine Ecology Progress Series* 131: 115–124.

Vathana, K. and P. Penh. 2003. *Review of Wetland and Aquatic Ecosystems in the Lower Mekong River Basin of Cambodia.* Phnom Penh: Cambodia National Mekong Committee.

Walters, B. B., P. Rönnbäck, J. M. Kovacs, B. Crona, S. A. Hussain, R. Badola, J. H. Primavera, E. Barbier, and F. Dahdouh-Guebas. 2008. Ethnobiology, socio-economics and management of mangrove forests: A review. *Aquatic Botany* 89: 220–236.

Chapter 8

Rosenfield, P. L. 1992. The potential of transdisciplinary research for sustaining and extending linkages between the health and social sciences. *Social Science & Medicine* 35(11): 1343–1357.

Index